Digital Zoology
Version 1.0
Student Workbook
and
CD-ROM

Jon G. Houseman
University of Ottawa

Boston Burr Ridge, IL Dubuque, IA Madison, WI New York San Francisco St. Louis
Bangkok Bogotá Caracas Kuala Lumpur Lisbon London Madrid Mexico City
Milan Montreal New Delhi Santiago Seoul Singapore Sydney Taipei Toronto

Quickstart Information

Windows 95, 98, NT, ME, 2000

***Running the Digital Zoology* Program**

- Insert the CD into your computer's CD-ROM drive. Auto-start has been implemented with this CD so the *Digital Zoology* Program will automatically start at this point for most users.
- If Auto-start does not start the *Digital Zoology* program, double-click on the **tree.exe** file located at the top level of the CD.

Macintosh Power PC, OS 8.5 or Later

Running the *Digital Zoology* Program

- Insert the CD into your computer's CD-ROM drive. Double-click on the CD icon if the program does not auto-start.

McGraw-Hill Higher Education

A Division of The McGraw-Hill Companies

DIGITAL ZOOLOGY VERSION 1.0
CD-ROM AND STUDENT WORKBOOK

JON G. HOUSEMAN

Published by McGraw-Hill Higher Education, an imprint of The McGraw-Hill Companies, Inc., 1221 Avenue of the Americas, New York, NY 10020.

This book is printed on recycled, acid-free paper containing 10% postconsumer waste.

1 2 3 4 5 6 7 8 9 0 QPD/QPD 0 3 2 1
ISBN 007-248950-2

www.mhhe.com

TABLE OF CONTENTS

AUTHOR'S PREFACE

About *Digital Zoology*

Digital Zoology has been a variety of different things over the years. It started as a series of digital photos that I used in class to help bring the specimens that students saw in lab to the lecture hall. It grew with the addition of more specimens and interactive cladograms. Soon after that it found a home on the computer networks at the University. When students wanted to take it home, it ultimately became the CD-ROM that evolved into the *Digital Zoology Free Preview Version*, which published in December, 2000. By January 2001 over 15,000 students across North America were using the program. Now, I have the pleasure of welcoming you to *Digital Zoology* Version 1.0. Version 1 includes some major changes and you are holding the most obvious-the Student Workbook. The Workbook provides a quick look at different animals traditionally studied in zoology labs and serves as a useful study tool for review of lab dissection exercises.

The Student Workbook

The Student Workbook is designed to complement the specimens on the *Digital Zoology* CD-ROM. If a particular taxon isn't included in the current version of *Digital Zoology,* there won't be a chapter on it in the workbook. Each chapter begins with a short description of the defining differences for the group of organisms — phylum or class — covered in the chapter. These differences, many of which can be seen in the lab or on the *Digital Zoology* CD-ROM, combine characteristics that define the taxon and highlight the unique ways that organisms in the group function. The Workbook takes a closer look at a representative specimen for each taxon and explores the unique biology of the animal. A convenient "structures checklist" allows students to keep track of the key structures for each of the specimens they examine. Each chapter contains self-evaluation tools: crossword puzzles that test knowledge of key terms and concepts for that group, and a labeling quiz for lab exercises. For those of you who have seen the Digital Zoology Free Preview Version, the labeling quizzes in this Workbook are similar to the drag-and-drop quizzes on the Digital Zoology CD-ROM. They are still on the Version 1 CD-ROM, only there are many more of them: a quiz is now included for all the major animal groups covered in *Digital Zoology.* Some of the additions to Version 1 include:

- new vertebrates, including the fetal pig, perch, and frog;
- expanded interactive glossary;
- a cladogram for the pseudocoelomates; and
- specimen lists to make it easier to find the different organisms covered in *Digital Zoology.*

There are a number of other interactive elements in the Student Workbook. Questions in each chapter explore some of the key concepts about the group of animals discussed, and the *Digital Zoology* web site provides answers to the questions raised in the Workbook. Crossword puzzles provide a self-evaluation tool for each chapter's content and on-line interactive versions of the puzzles, with solutions and hints, are available on the web site. An additional interactive element in each chapter of the Workbook is the Self-Test quiz. Students can identify and label key structures in photographs of microscope slides or dissected specimens.

This Workbook is not intended to replace the zoology textbook or laboratory manual. Rather, it has been designed as a supplement to both. In a large part, *Digital Zoology* focuses on things that students will be able to see in their lab sessions.

A full color copy of the workbook is available as an Adobe Acrobat file in the Workbook folder on the *Digital Zoology* CD-ROM.

The CD-ROM

Digital Zoology CD-ROM is organized around two components. The first provides illustrations, photos, and video clips of the various specimens that will help to understand the specimens and groups of animals that are commonly studied in the lab. The second involves interactive cladograms that explore the origins of the various animal taxa at the phylum, subphylum and class level.

The *Digital Zoology* CD-ROM contains over 35 different species, covering a full range of organisms from microscopic amoebas to vertebrates such as the frog and fetal pig. It provides students in whole animal biology courses such as General Zoology, Invertebrate Zoology, and Vertebrate Zoology with an interactive guide to the specimens and materials that they will be studying in their laboratory and lecture sessions. Each lab module contains illustrations, photographs, and annotations of the major structures of organisms and microscope slides commercially available from the suppliers used by high schools and universities. Lab modules are combined with explanations of the various animal groups and interactive cladograms that allow students to investigate the major evolutionary events that have given rise to the tremendous diversity of animals that we find on the planet. Each lab module is accompanied by a drag-and-drop quiz that allows students to test their understanding of the specimens that they have been looking at.

A Note to Students

I hope that you'll find *Digital Zoology* to be a valuable tool in your studies of zoology. Here are some of the ways that my students have used the program.

Lab Preparation. *Digital Zoology* is a great way to prepare for lab sessions. Part of this preparation should always be to read through the lab manual before your lab. But, often manuals have only a few illustrations or pictures of the specimens you'll be looking at in the lab. They really don't prepare you for how things will look when you get inside your own specimen. *Digital Zoology* contains lots of pictures of the specimens at various stages in the dissection and by working through the lab module prior to actually doing the lab you should be able to get a good idea of what to expect once you start your own dissection.

Almost all lab courses have some sort of lab quiz where you have to look at a specimen and answer some question about it (for example: "identify the structure labeled A and what it does"). I've tried to duplicate that with the printed Self-Test here in the Student Workbook and the drag-and drop quizzes on the *Digital Zoology* CD-ROM. Solutions to the printed Self-Tests are available on the *Digital Zoology* web site, and the drag-and-drop quizzes on the *Digital Zoology* CD-ROM are interactive and allow you to find the correct answers. If you want even more practice at identifying structures, place a sticky note over the lower right corner of the screen to hide the legend in *Digital Zoology*. Take out a piece of paper and write what you think the legend should be, using the lettered pointers on the screen. Compare your legend to the one in *Digital Zoology*.

Learning the Language. Zoology, like any discipline, has a set of terms, or lexicon, all its own. The text in the CD-ROM version may appear to be full of words that don't have any meaning to you yet as you start out in your course, but you'll know the terms by the end of it. In the *Digital Zoology* CD-ROM, key words and terms are highlighted in blue. As you read the text in the taxon information boxes and within the interactive cladograms, resist the urge to click on a highlighted term right away. Instead, take a scrap of paper and write out a short definition for the term. Then, click on the term and compare your definition to the one in *Digital Zoology*. A full glossary is available in the Student Workbook.

One of the biggest hits with my students are crossword puzzles, and I've provided paper versions in the Workbook and on-line interactive versions on the *Digital Zoology* web site. The clues in crossword puzzles are similar to the type of one-word answers, or fill-in-the-blank type of questions that often appear on exams and quizzes. Each puzzle has between 30 and 40 different clues to test your knowledge on each of the different groups of animals.

Finally, throughout out the Workbook you'll find some short answer questions on animals being discussed in that chapter. You'll find answers to these questions on the *Digital Zoology* web site.

A Note to Instructors

Many of the photographs and illustrations on the CD-ROM version of *Digital Zoology* are available for you to use in your own class. You can download PowerPoint slides from the *Digital Zoology* web site (www.mhhe.com/digitalzoology/instructors) for each of the major taxa covered in *Digital Zoology*. Feel free to modify, alter, or customize the slides to meet your own requirements. The design and color scheme of the slides has been optimized so the images can be used as 35mm slides, transparencies, or, if you've access to it, digital projection. If you find a photograph or illustration in *Digital Zoology* that isn't included in the PowerPoint slides, all of the pictures on the CD-ROM are available on the web.

The images used in *Digital Zoology* are part of the large BIODIDAC image base (biodidac.bio.uottawa.ca) created for biology teachers to use in their courses. BIODIDAC is funded in part by Heritage Canada, a ministry of the Government of Canada. The image bank consists of over 3,000 photographs, illustrations, annotated drawings and video clips of biological materials. All the material in the BIODIDAC image bank has been donated by biology teachers and the donated materials are available royalty and copyright-free for the creation of non-commercial teaching materials. BIODIDAC is always looking for additional materials, and if you would like to contribute you can find out how by visiting BIODIDAC.

I'd like to take this opportunity to share with you some observations on how *Digital Zoology* has changed my lectures and labs over the past 10 years of its use. The ability to bring photos of the same materials students see in the lab to the lecture provides a clear link to bind these two components of the course that much closer together. Our analysis of student use of *Digital Zoology* has shown us that most of the students go through the lab module prior to actually doing the lab. The students tell me that the photos help them to anticipate what their dissection is going to look like during various stages in the dissection. We actually have a few computers in the lab with *Digital Zoology* running, and as students dissect they look for a visual concordance between their own work and the CD-ROM modules. *Digital Zoology* wasn't designed to replace the dissection of specimens, it was my hope that students would gain a better understanding of the dissection through the use of the program. The crossword puzzles have been a tremendous success with the students. The clues in the puzzles are similar to fill-in-the-blank questions, and all the clues are available at the instructors' web site for *Digital Zoology*.

A Selected Bibliography

The following resources were invaluable in preparing the *Digital Zoology* Student Workbook and CD-ROM.

Brusca, R.C. and Brusca, G.J. *Invertebrates*. Sinauer Associates Inc.

Hickman, C.P., Roberts, L.S. and Larson., A. 2001. *Integrated Principles of Zoology*. 11th edition. McGraw-Hill Higher Education.

Kardong, K.V. 1997. *Vertebrates: Comparative Anatomy, Function, Evolution*. 2nd edition. WCB/McGraw-Hill.

Moyle, P.B. and Cech, J.J. 2000. *Fishes: An Introduction to Ichthyology*. 4th edition. Prentice Hall.

Nielsen, C. 1995. *Animal Evolution: Interrelationships of the Living Phyla*. Oxford University Press.

Pechenik, J.A. 2000. *Biology of the Invertebrates.* 4th edition. McGraw-Hill Higher Education.

Pough, F.H., Andrews, R.M., Cadle, J.E., Crump, M.L., Savitzky, A.H. and Wells, K.D. 2001. *Herpetology*. Prentice Hall.

Walker, W.F. and Liem, K.F. 1994. *Functional Anatomy of the Vertebrates: An Evolutionary Perspective.* Saunders College Publishing.

Willmer, P. 1990. *Invertebrate Relationships*. Cambridge University Press.

A Special Thank-you

I'd like to especially thank the past, present and future students of my Animal Form and Function and Invertebrate Zoology courses here at the University of Ottawa. They have been tremendously patient with my adventures in the use of these new technologies in teaching and after 10 years I, and they, are convinced the adventure was well worth it. Thanks to Anne Scroggin for her copy editing of the Workbook and thanks to the folks at McGraw-Hill who have helped to make this all happen: Marge Kemp, Donna Nemmers, Dianne Berning, Linda Avenarius, Audrey Reiter, and Mark Christianson, constantly manage to pull a rabbit out of the hat exactly when it's needed. Also, thank you to the many users of the *Digital Zoology Free Preview Version* who provided useful comments to us through the *Digital Zoology* web site. It is our goal to make each version of *Digital Zoology* more expansive and exciting than the versions before, and your suggestions help us reach that goal.

As you progress through *Digital Zoology* Version 1.0, please let me know your comments. If there is something I can do better, or something that you would like to see added to the program, let me know. You can get in touch by using the *Digital Zoology* web site at www.mhhe.com/digitalzoology.

— Jon G. Houseman
May 2001

Digital Zoology CD-ROM Components

When the *Digital Zoology CD-ROM* first opens, a dendogram (tree) of the major animal and protozoan phyla appears on the screen (fig. 1). The cladogram button on the lower right of the title bar gives a cladistic representation of the same material displayed, with a dendogram button to return to the initial screen.

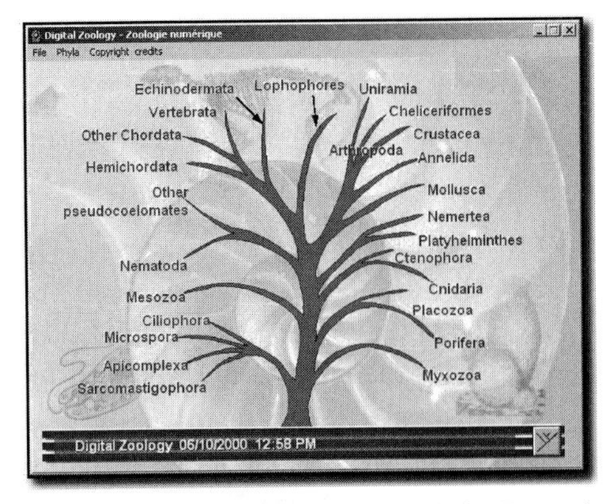

Figure 1: Main dendogram (tree) for *Digital Zoology CD-ROM*

Taxon information box

In either view, selecting the name of one of the taxa opens the taxon information box containing information about that group, along with a randomly selected picture illustrating animals from that group (fig. 2). A scrolling text box includes information on the numbers, relative importance, and unique aspects of how that taxon functions. Blue underlined words in the taxon information box's scrolling dialogue are hyperlinked to an interactive glossary (fig. 2). The bottom of the taxon box includes buttons that open a cladogram and also lab modules for that taxon. If there is more than one lab module for the taxon, you will need to open the cladogram to access the labs. If the button is "grayed out," there is no additional information available.

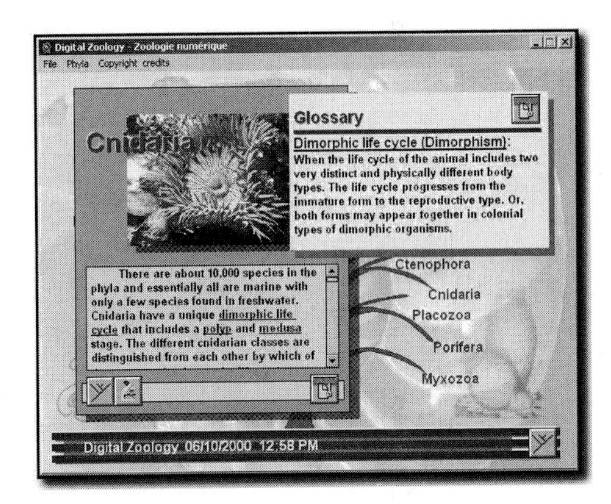

Figure 2: Taxon information box and the interactive glossary.

Interactive cladograms

The cladogram button appears on the main tree in the lower right-hand corner and also in the taxon information boxes. When you click on the button for the main tree, a cladistic diagram of the major animal phyla appears. Using the cladogram button within a taxon information box will display a cladogram for the classes (or other subordinate taxon) for that phylum. In both cladograms, interactive synapomorphies are represented by question marks that, when activated by the mouse click, open a text box explaining the main evolutionary events at that point (fig. 3). Here again, access to the interactive glossary is available through words that appear underlined and in blue.

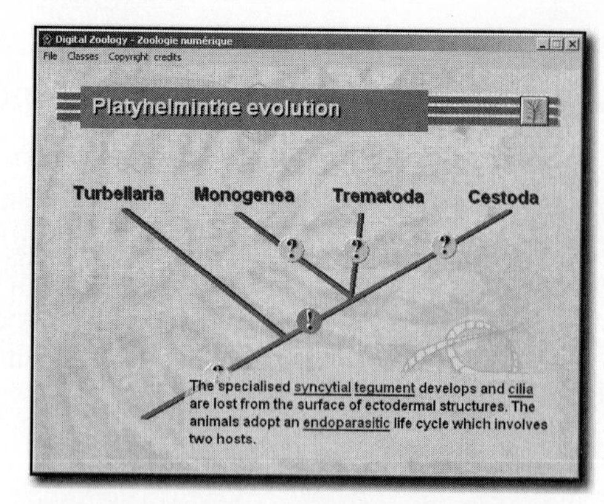

Figure 3: Interactive cladogram showing the synapomorphic characters that define the different groups within the taxon.

Laboratory modules

Laboratory modules consist of different topics, further divided into subtopics. For example, the topic "internal anatomy" might contain subtopics on "general features," "anterior structures," and "posterior structures." Or, a phylum may be the topic, with subtopics for different animals within it. The topic list for each lab module is always available on the menu bar at the top of the screen (fig. 4), and the list of subtopics appears to the right of it. For each of these subtopics, the figure navigation bar notes the number of figures. Its right and left arrows move you forward and backward through the figures. The number of subtopic pages is shown in the subtopic navigation bar, and its right and left arrows can help you navigate through the different subtopic pages that are available.

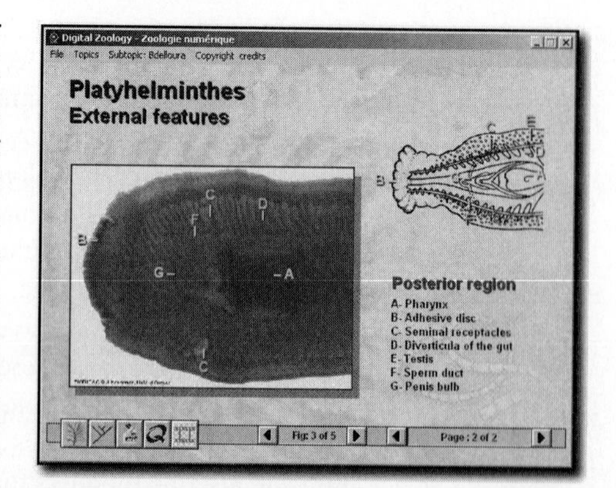

Figure 4: Lab module screen. The navigation bar at the bottom has from left to right: navigation buttons, figure navigation, and subtopic page navigation.

Drag-and-drop quizzes

A number of the laboratory modules have Drag-and-drop quizzes that test your familiarity with material in that module (fig. 5). Use the quiz button found on the module's navigation bar to start the quiz. The legend for each of the pinheads is given in the list, and below it a set of pinheads with labels is provided. Click and hold the left mouse button down over the labeled pinhead, and drag it into the circle that corresponds to the structure identified in the legend. To change your answer, drag the lettered pinhead off the circle. To check out your identifications, use the "check" button to correct the quiz. Green pinheads appear to show which of your answers was correct, and red pin heads show you the structures that you didn't label properly. Use the right arrow on the question navigation bar to move to the next question. Each time the quiz runs, the questions are slightly different—you might want to try it more than once! There are other ways that you can use *Digital Zoology CD-ROM* as a study tool. To test how well you recognize different structures in the lab modules, cover the legend that appears on the lower right corner of the screen with a piece of paper. Write a legend based on the labels in the picture and then compare your answers to the original. You can also use the *Digital Zoology CD-ROM* to test your familiarity with key zoological terms. When you read through any of the explanations and see a term highlighted in blue, write a short definition. Then click on the word and compare your definition to the one that appears on the screen.

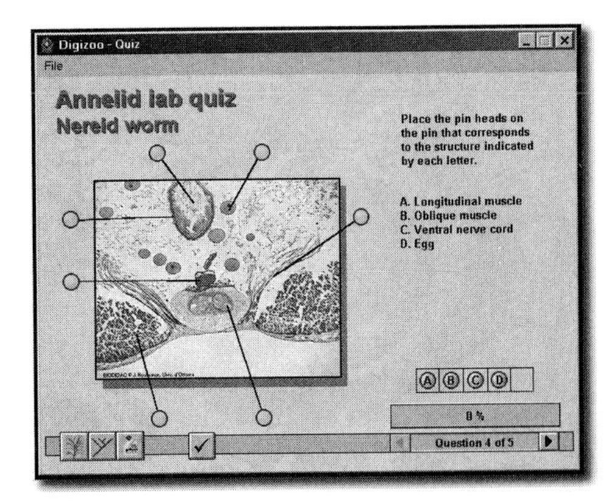

Figure 5: Drag-and-Drop quiz module.

Navigation

Digital Zoology CD-ROM uses navigation in the form of buttons, menu bars, and keyboard shortcuts. Use the following list to find out what each button does, or place the mouse pointer over any of the buttons, wait a second, and an explanation of what the button does will appear on the screen.

Buttons

Cladogram Button: Click this button to display the cladogram for the taxon that has been selected. If a phylum's cladogram button is clicked, the cladogram for the classes in that phylum opens. On the main tree, this button switches to a cladogram version of the same material.

Dendogram Button: Click this button to open the main tree from anywhere on *Digital Zoology CD-ROM.*

Close Button: Click on this button to close the taxon information box or the glossary dialogue box (shortcut—ESC).

Lab Button: When this appears in a taxon information box, it will take you to the corresponding lab module. When this button appears in a lab module, it takes you to the start of the module (shortcut—HOME).

Quiz Button: The quiz button is found in the lab manuals and will take you to an interactive quiz that tests your knowledge on the material in that lab.

Movie Button: In some parts of *Digital Zoology CD-ROM*, video clips are available. Click this button to start and stop the movie.

Question Button: Click this button on a cladogram for an explanation of the synapomorphies for that point in the cladogram.

Explanation Button: You can't click on this one in a cladogram, but it will tell you the point on the cladogram to which the explanation applies.

Previous Subtopic Page or Figure: Lab modules are divided into topics, which contain subtopics that in turn have figures. Using this button either moves you to the previous subtopic page of the topic (shortcut—Down arrow) or the figure on that page (shortcut—Left arrow).

Next Subtopic Page or Figure: Lab modules are divided into topics that contain subtopics, which contain pages that in turn have figures. Using this button either moves you to the next page of the topic (shortcut—Up arrow) or the figure on that page (shortcut—Right arrow).

Menus

The main menu bar at the top of each module changes depending where the user is in the program. The first item on the left is always "File" with two choices: "File: Back" and "File: Exit Digital Zoology." "File: Back" closes the module currently open and returns the user to the module that had previously been opened (the dendogram button is always available to return to the main tree on the opening screen). "File: Exit Digital Zoology" exits the program and closes all the modules.

What appears in the second position of the menu bar depends on where the user is in the program. In taxonomic modules (either cladograms or dendograms), the second item on the menu bar is either a list of the taxon modules—either phyla or classes—that the user can open or it contains a list of other specimens from the same taxon. These correspond to the linked names that appear on the dendogram or cladogram. In Lab modules, the next item on the main menu bar lists the major topics available in the module, and this corresponds to the links that appear on the screen.

In 3ab modules, once a topic is chosen, a list of the different subtopics appears in the third position on the menu bar. Each page has a variety of figures associated with it.

Digital Zoology Version 1.0 CD-ROM Correlation Chart

	Digital Zoology CD-ROM Resources*	Hickman et al., *Integrated Principles of Zoology*, 11th ed.	Hickman et al., *Animal Diversity*, 2nd ed.	Miller & Harley, *Zoology*, 5th ed. ‡web sites	Hickman & Kats, *Laboratory Studies to Accompany Integrated Principles of Zoology*, 13th ed.	Lytle, *General Zoology Laboratory Guide*, 13th ed.	Pechenik, *Biology of the Invertebrates*, 4th ed.
Protozoa: Sarcomastigophora	B,P,V	11	4	8	6	5	3
Ciliophora	B,P,V	11	4	8	6	5	3
Apicomplexa	B,P	11	4	8	6	5	3
Microspora	B	—	—	8	—	—	3
Myxozoa	B	—	—	8	—	—	3
Placozoa	B	12	—	9‡	—	—	4
Mesozoa	B	12	—	9‡	—	—	9
Porifera	B,P,Q	12	5	9	7	6	4
Cnidaria	B,C,P,Q	13	6	9	8	7	6
Ctenophora	B	13	6	9	—	—	7
Platyhelminthes	B,C,P,Q,V	14	7	10	9	9	8
Nemertea	B	14	7	10	—	—	10
Nematodes & Pseudocoelomates	B,C,P,Q,V	15	8	11	10	10	11,16
Mollusca	B,C,P,Q	16	9	12	11	11	12
Annelida	B,C,P,Q,V	17	10	13	12	12	13
Arthropoda: Chelicerata	B,C,P	18	11	14	13	13	14
Crustacea	B,C,P,Q	19	11	14	14	13	14
Uniramia	B,C,P,Q,V	20	11	15	15	13	14
Lophophorates	B,C	22	12	16‡	—	—	19
Echinodermata	B,C,P,Q	23	13	16	16	14	20
Hemichordata	B,C,P	24	13	17	—	—	21
Protochordates	B,C,P	25	14	17	17	15	22
Vertebrata: Agnatha	B,P	26	15	18	18	—	
Chondrichthyes	B,P,Q	26	15	18	18	16	
Osteichthyes	B,P,Q	26	15	18	18	17	
Amphibia	B,P,Q	27	16	19	19	18	
Reptilia	B	28	17	20	20	—	
Aves	B	29	18	21	21	—	
Mammalia	B,P,Q	30	19	22	22	19,20	

* B=basic text, C=cladogram, P=pictures of lab materials, Q=quiz, V=video

PROTOZOANS

Inside *Digital Zoology*

As you explore the protozoans on the *Digital Zoology* CD-ROM, don't miss these learning tools:

Photos of prepared slides from a wide variety of protozoans including: amoebas (*Amoeba*, *Difflugia,* and *Actinosphaerum*); flagellates (*Euglena*, *Trypanosoma*, and a variety of Phytomastigophora including *Volvox*); ciliates (*Paramecium*, *Blepharisma*, *Stentor*, *Spirostomum*, *Didinium*, *Vorticella*, and *Euplotes*, among others); and the causative organism of malaria, *Plamodium,* as an example of Apicomplexa (Sporozoa)

Video clips of amoebas, flagellates, and ciliates give a you a good look at how these organisms move. You can also watch water expulsion vesicles empty in *Paramecium* and *Stentor*, see metachronal waves in a number of species, watch how *Stentor* feeds, and see the rapid escape reaction of *Vorticella*.

Defining Differences

Some of the differences described in the following sections appear for the first time in the protozoans, and they define the group. Others are important for understanding how these organisms function. Whichever is the case, you'll want to watch for examples of each in the various protozoans that you will be examining.

- Unicellular "animals" with
- specialized organelles including
- cilia, flagella, or pseudopodia used for locomotion.

Unicellular "animals"

The single-cell organisms, protists, have always been a problem for taxonomists, who have trouble defining categories for the organisms. Their mixed feeding habits, which include being capable of photosynthesis — **autotrophic**, ingestion of other organisms — **heterotrophic**, or survival on dead and decaying materials — **saprophytic**, makes them single-cell versions of the plant, animal and fungal kingdoms. The kingdom Protista probably shouldn't be a kingdom on its own. Instead, single-celled **eukaryotes** may be better placed in the other biological kingdoms. In a zoology course the protists that we are most interested in have animal-like traits: no cell wall, heterotrophic, and capable of movement — the protozoans.

Protozoans are microscopic in size, although some are large enough to be seen with the naked eye. Microscopic organisms such as these have a number of advantages, one of which is that they have a high surface-to-volume ratio. That's an advantage because diffusion is an important process for many of their functions. Don't let their small size fool you; these are complex organisms. Everything they do, and how they do it, is packed into a single cell. In animals, organs and tissues become specialized for certain functions; for protozoans, it's their organelles that take on the same functions.

Like all animal cells, protozoans are covered by a **plasma membrane** that surrounds the **cytoplasm** of the cell, the protozoans integument or skin. Like all membranes, it's permeable; water and low molecular weight substances dissolve in it. Oxygen, required for metabolism, and metabolic wastes generated by metabolism, diffuse across the surface, giving the plasma membrane respiratory and excretory functions. The **nucleus,** with its **nucleolus** and **chromosomes,** contains the genetic material and will be involved in providing copies of itself for new cells, the reproductive system. The **golgi apparatus, rough endoplasmic reticulum,** and the **lysosomes** that they produce, fuse with ingested **food vacuoles, phagosomes.** In some protozoans, food is always ingested at the same place on the cell surface, the **cytostome** (cell mouth). Undigested food is removed at a **cytoproct**, or the cell anus. All of this is similar to a digestive system. **Mitochondria**, which fuel life's processes in all cells, use the nutrients from digested food and oxygen that has diffused across the cell surface. In freshwater protozoans the **water expulsion vesicle** is an osmoregulatory organelle removing excess water that diffuses into the cell. Protozoans have **endoskeletons** formed from the microtubular **cytoskeleton** of the cytoplasm or **exoskeletons** secreted as **shells** or **tests**. Some of these single-cell organisms are capable of defending themselves using organelles such as **trichocysts,** and others can orient to external stimuli and detect light with **stigmata**. Movement is even controlled by organelles, **cilia** and **flagella**. These and the pseudopodia in the amoebas give these small creatures movement, something we associate with animals. Essentially, everything that a large animal can do, single-cell protozoans can do as well, only on a much smaller scale.

Specialized organelles

You won't be able to see all these different organelles with the light microscope, so we'll focus on those you'll probably see in your lab or *Digital Zoology*.

The **nucleus** contains the genetic information of the cell. In addition to activities that maintain the cell, the nucleus creates copies of itself for new protozoans by either **asexual reproduction** — using mitosis, or **sexual reproduction** — using **meiosis.** Asexual reproduction often involves **fission**, or some sort of division of the parent cell into daughter cells. If the two daughters are similar in size, it's **binary fission**. If one's large than the other, it's **budding**. If there has been multiple mitotic divisions and the parent cell divides into more than two daughters, it's **multiple fission**. Some protozoans form gametes that fuse to form a new protozoan. Ciliates have two nuclei: a **macronucleus** for running the day-to-day functions of the cell and a **micronucleus** for reproduction. Ciliates have a form of genetic mixing, **conjugation**, not found in the other protozoans.

✍ *What are the similarities and differences between sexual reproduction and conjugation?*

The **water expulsion vesicle** (also called a **contractile vesicle**) is an important osmoregulatory organelle in freshwater protists, which are **hyperosmotic** to the surrounding water. Water constantly diffuses into the cell and could cause the cell to burst if there wasn't some way to get it out. That is what the water expulsion vesicle does. A network of **collecting tubules** extends throughout the cells, and the rich supply of mitochondria on the surface of the tubules provides the energy needed to pump water from the cytoplasm into the tubules. The tubules dump the water they've collected into the water expulsion vesicle, which slowly fills. Once full, the vesicle's membrane fuses with the outer plasma membrane creating a pore out of which the excess water flows.

Protozoans ingest food by **phagocytosis**, and as a result, the ingested food is surrounded by a plasma membrane creating a **food vacuole**, or **phagosome**. Digestion doesn't start until the phagosome fuses with a **lysosome** that contains the digestive enzymes. Everything being digested, and the enzymes digesting it, are contained inside a plasma membrane permeable only to small molecular weight compounds, the products of digestion. The enzymes are too big to cross the membrane, so there's no danger that they might mistakenly digest the cytoplasm of the cell. Initially the pH inside the food vacuole is acidic and as digestion proceeds, the contents of the vacuole become more alkaline. To be sure that nutrients get to every part of the cell, the food vacuole is swept around the cell by the cytoplasmic currents and nutrients diffuse out supplying the different parts of the cell. Once there is only undigested food left, the food vacuole fuses with the cell membrane, dumping its contents to the outside of the cell.

Cilia, flagella, and pseudopodia

Cilia and **flagella** share a common structure, with a microtubular core that has a 9+2 organization with nine paired doublets of **microtubules** surrounding a central pair to form the **axoneme**. The axoneme is surrounded by plasma membrane and anchored to the inside of the cell by the **kinetosome** formed from nine microtubular triplets connected to the nine doublets in the axoneme. Inside the axoneme arms, the nine doublets extend arms to their neighbors, and when ATP is burned, the position of these arms changes and the microtubules slide past each other. That sliding motion is what bends the axoneme and in turn, the cilium or flagellum bends.

What are the differences between cilia and flagella?

In protozoans flagella are always found one per cell. Cilia are much more numerous and, depending on where on the cell surface they are found, they can be modified for different functions. Cilia can fuse to make even bigger structures, creating complex ciliature such as **cirri** and **membranelles**.

Cilia and flagella can generate substantial force as they try to pull or push a protozoan through the water. To be able to do that, the plasma membrane needs to be strengthened, and when it is, the surface is referred to as the **pellicle**. In some ciliates even further strengthening occurs with the **kinetosomes** being anchored to each other creating a complex **infraciliature**. The stronger the pellicle, the less likely it is that the shape of the protist can change, and most ciliates have recognizable shapes. A good example is *Paramecium* with its oval shape and food groove.

Movement in a pseudopod involves changes in the fluidity of the cytoplasm from the more solid **plasmagel (ectoplasm)** or fluid **plasmasol (endoplasm)**. Like muscles in higher animals, pseudopodial movement involves actin, mysosin, and ATP.

Describe the biochemical events involved in changing endoplasm to ectoplasm and back to endoplasm.

A Closer Look Inside *Digital Zoology* - Protozoans

Prepared slides and wet mounts of live specimens are the best way to examine the protozoans. Internal structures are easier to see in prepared slides because histological stains highlight nuclear and proteinaceous structures. Prepared slides don't, however, give a sense for how the organisms move — that's the value of the wet mount. If you can, try to look at the two preparations together. Watch your specimen move, then look for the internal structures. If live and prepared specimens aren't available, you'll find the combination in the photos and videos in *Digital Zoology*.

Sarcodina: Amoebas

Sarcodine protozoans are amoebas, and there are two types: naked and shelled. It is the naked amoebas with which you might already be familiar.

As you examine the prepared slides you should be able to see the **water expulsion vesicle**, **nucleus**, **food vacuoles**, and **pseudopods**. In a wet mount some of these organelles will be visible, but with this preparation watch the pseudopods move. At the tip is the clear zone, the **hyaline cap**, that the fluid endoplasm pushes against before flowing to the edge of the pseudopod and turning into ectoplasm. Look at the other end of the amoeba and you'll see the reverse, ectoplasm turning back to endoplasm then flowing into the center of the advancing pseudopod. If you don't get a good view of it in your lab, check in *Digital Zoology*.

✍ *How many different types of pseudopods can amoebas have?*

The tests of shelled amoebas are made from a variety of substances including chitin, sand grains and substrate, calcium salts, and even silica. These shells still accumulate in the ocean depths and are referred to as strews. One of the more famous fossil strews are the white cliffs of Dover. The chalk from which they are made are the shells of ancient amoebas. This tells you a little about the diversity and numbers of these single-cell organisms that flourished in the oceans so long ago. Be sure to take a look at some of these shelled amoebas, especially the delicate silicaceous shells of the radiolarians.

Mastigophorans: Flagellates

As a rule, if a protist has only one flagellum, it's probably a **zoomastigophoran**. If it's a photosynthetic flagellate, a **phytomastigophoran** (one of the many types of algae), it will have two flagella. We've included both in *Digital Zoology* so you can compare the two.

In zoology courses the flagellate *Euglena* is often used for an example of a zoomastigophoran. If you set up your microscope properly, with not too much light shining on the specimen, you'll be able to see the **flagella** and **stigmata**, or **eye spot**, that the species uses to orient toward light. *Euglena* sits on the fence between being a unicellular animal and/or plant. It is capable of photosynthesis, that's why it's green; but if there is no light, it can survive as a **heterotroph**. Like ciliates, *Euglena* has a **pellicle** to anchor the flagellum to the cell. *Euglena*'s pellicle isn't as rigid as a ciliate's; that allows for the euglenoid movement that you may get a chance to see. If not, there's a video of it in *Digital Zoology*. Another flagellate example is *Trypanosoma,* which causes sleeping sickness. The trypanosome is found between the red blood cells. The base of its flagellum is anchored at the anterior end and is fused along the length of the cell in an undulating membrane that looks almost like a fin.

 *Describe the life cycle of the causative organism of sleeping sickness,* Trypanosoma cruzi.

Volvox is an example of a colonial phytomastigophoran, and it's often included in zoology courses because some believe it gives clues to how multicellular **Metazoa** first appeared. Many metazoans go through a hollow ball stage, the **blastula**, during their embryology. The assumption is that colonial *Volvox* may be some sort of prototype for that stage. **Spherical** and **radial symmetries** are often also thought to be **primitive traits** in the ancestry of the metazoa. If that's the case, again *Volvox* may provide another clue to metazoan origins. (Symmetry is discussed in more detail in the cnidarian chapter.) There are problems with this, not the least of which is that *Volvox* is a colony of diflagellate, photosynthetic algae, with **haploid** somatic cells. These traits are never seen in animals!

Ciliophora: Ciliates

One of the most often used ciliates in zoology courses is *Paramecium,* and no doubt you'll get to see it and a few others in your lab session. *Digital Zoology* includes *Paramecium, Blepharisma, Stentor, Spirostomum, Didinium, Vorticella, Euplotes*, along with a quick look at a few others.

We're looking at ciliates, so one of the things that you want to be sure to see are the cilia that cover the cell. Cilia and their **kinetosomes** are anchored into the plasma membrane with a reinforcing **pellicle,** and it's the reason why many ciliophorans have a consistent shape, or look. You can see it on the specimen as lines that appear to run the length of the organism. *Blepharisma* and *Spirostomium* are good examples for seeing the pellicle.

During **phagocytosis** the cell membrane surround the food and without disrupting the reinforcing pellicle. The solution is to always ingest food at the same place on the cell surface, the **cytostome**. The other side of the coin is that undigested food is also removed from the ciliate at a **cytoproct**. In the simplest ciliate, the ciliature is the same all over the body; in other's it becomes specialized. With feeding always happening at the same place on the cell surface the cilia in that area become modified for capturing food, and called **oral ciliature. Somatic cilia** cover the rest of the body and involved in locomotion. Good examples of modified cilia and the cytostome can be found in *Paramecium, Blepharisma, Stentor,* and *Vorticella.*

Cilia on the cell surface move in a coordinated manner. Not all of the cilia are at the same point in the power and recovery stroke, and that appears as a **metachronal wave.** Although you can see it on a number of the specimens in *Digital Zoology*, the wave that runs around the lip of *Stentor* is a good example. While some swim using cilia, other ciliophorans remain attached to the substrate — *Stentor* and *Vorticella,* for example. Both specimens have contractile elements and are able to move using these; watch *Voriticella* with the coiled stalk and its internal **myoneme**.

The macronucleus is easy to see in the prepared slides and the living specimens. If you don't get to see a water expulsion vesicle empty during the lab, be sure to see it with *Paramecium* and *Vorticella* in *Digital Zoology*.

Apicomplexa: Plasmodium

The causative agent of malaria, *Plasmodium,* is an example of an Apicomplexan (Sporozoan). Like all **parasites**, what a disease organism is, this protozoan has a complex life cycle. This complexity is, in part, so the parasite can move from one host to another. (Parasitism is discussed in more detail in the platyhelminth chapter.) Malaria moves from mosquitoes to humans and back. If the different stages of the life cycle aren't available, you'll find pictures of sporozoites, merozoites, trophozoites (inside the red blood cells), and oocysts in the gut wall of the mosquito in *Digital Zoology*.

At one time it was thought that the world was free of malaria because the mosquito that carried it could be controlled with pesticides, which at that time included DDT. Unfortunately the announcement was premature. The mosquito survived and is now resistant to insecticide treatments. From only a few reported cases a year in the 1960s, malaria now kills more than three million people annually.

✍ *What are the different stages in the malaria's life cycle, and what is the functional purpose of each?*

Cross-References to Other McGraw-Hill Zoology Titles

Integrated Principles of Zoology, 11th edition. C.P. Hickman, L.S. Roberts & A. Larson. Chapter 11.

Animal Diversity, 2nd edition. C.P. Hickman, L.S. Roberts & A. Larson. Chapter 4.

Zoology, 5th edition. S.A. Miller & J.P. Harley. Chapter 8.

Biology of the Invertebrates, 4th edition. J. Pechenik. Chapter 3.

Laboratory Studies in Integrated Principles of Zoology, 10th edition. C.P. Hickman, F. Hickman & L. Kats. Chapter 6.

General Zoology Laboratory Guide, 13th edition. C. Lytle. Chapter 5.

Structures Checklist

Here are some of the structures that you should be able to easily find in *Digital Zoology* and the specimens that you will be looking at in your lab. After reading your lab handout, you might want to add more and, depending on the equipment available in your lab, you might see more. As you study the material, you might also want to make some notes on how some of these structures looked or include a drawing in your lab notes. (Structures indicated by * may be hard to see.)

Sarcodines - slides and/or wet mounts

Amoeba **(slide and wet mount)**

- ☐ Ectoplasm
- ☐ Endoplasm
- ☐ Food vacuole
- ☐ Hyaline cap
- ☐ Nucleus
- ☐ Plasmalemma
- ☐ Pseudopod (lobopod)
- ☐ Water expulsion vesicle
- ☐
- ☐

Difflugia

- ☐ Pseudopod
- ☐ Pylome
- ☐ Test (shell)

Actinospaerum

- ☐ Axopodia
- ☐ Cell body
- ☐

Additional structures

- ☐ Foraminiferan strew
- ☐ Radolarian strew

Mastigophorans - slides and/or wet mounts

Euglena

- ☐ Cell body
- ☐ Chloroplast
- ☐ Contractile vacuole*
- ☐ Flagellum*
- ☐ Nucleus*
- ☐ Paramylon body
- ☐ Pellicle
- ☐ Plasmalemma
- ☐ Stigma
- ☐
- ☐

Trypanosoma

- ☐ Flagellum
- ☐ Nucleus
- ☐ Plasmalemma
- ☐ Red blood cells (host)
- ☐ Undulating membrane
- ☐

Volvox

- ☐ Asexual colony
- ☐ Daughter colony
- ☐ Parent colony
- ☐ Protoplasmic strands
- ☐ Reproductive cells*
- ☐ Sexual colony
- ☐ Somatic cells
- ☐
- ☐

Additional structures

- ☐
- ☐
- ☐
- ☐
- ☐

Ciliates - slides and/or wet mounts

Paramecium

- ☐ Binary fission
- ☐ Cilia
- ☐ Conjugation
- ☐ Cytostome
- ☐ Food vacuole
- ☐ Macronucleus
- ☐ Micronucleus*
- ☐ Oral groove
- ☐ Pellicle
- ☐ Water expulsion vesicle
- ☐
- ☐

Blepharisma

- ☐ Cilia
- ☐ Complex ciliature
- ☐ Macronucleus
- ☐ Oral groove
- ☐ Pellicle
- ☐ Water expulsion vesicle
- ☐
- ☐

Stentor

- ☐ Basal attachment
- ☐ Cilia
- ☐ Complex ciliature
- ☐ Cytostome
- ☐ Food vacuole
- ☐ Gullet
- ☐ Macronucleus
- ☐ Pellicle
- ☐ Stalk
- ☐ Water expulsion vesicle
- ☐
- ☐
- ☐

Vorticella

- ☐ Cilia
- ☐ Complex ciliature
- ☐ Cytostome
- ☐ Food vacuole
- ☐ Macronucleus
- ☐ Pellicle
- ☐ Stalk
- ☐ Water expulsion vesicle
- ☐
- ☐

Additional observations

- ☐ Metachronal waves
- ☐ Cytoplasmic streaming
- ☐
- ☐
- ☐
- ☐

Malaria

Prepared slides

- ☐ Merozoites
- ☐ Oocysts
- ☐ Red blood cells (host)
- ☐ Sporozoites
- ☐ Trophozoites
- ☐
- ☐

Crossword Puzzle – Protozoans

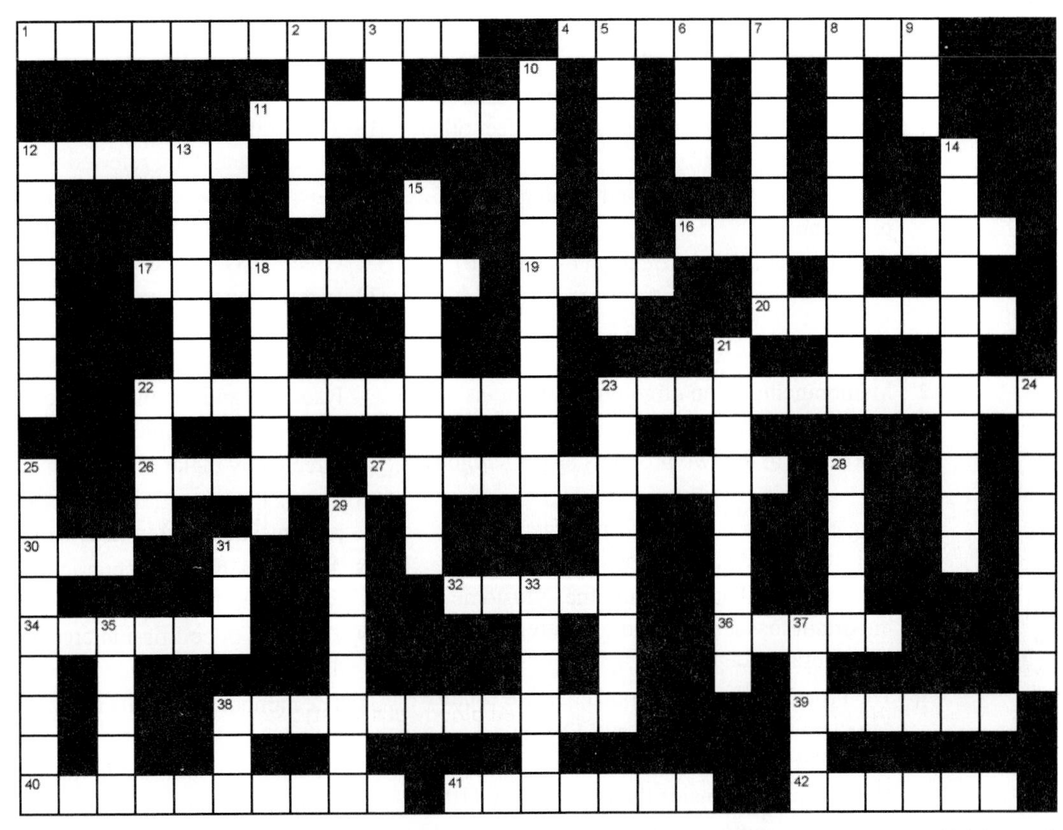

Puzzle solution. An interactive web based version of this puzzle, and its solution, are available on the *Digital Zoology* web site www.mhhe.com/DigitalZoology/Students. With the interactive puzzle you can check to see if individual words or the whole puzzle is correct, and get hints for single letters.

Across

1 During conjugation, this nucleus fuses to form the synkaryon. (12)

4 The more solid of the two protoplasm states in an amoeba. (10)

11 Mastigopharans use these for locomotion. (8)

12 Apicomplexan protists have this type of complex that defines the group. (6)

16 In apicomplexan protists, sporozoites result from this reproductive process. (9)

17 When a larger egg fuses with a smaller sperm, this combination of gametes is referred to as this. (9)

19 The internal structure of a cilia, or flagella, is referred to as a _____ plus two organization of microtubules. (4)

20 Once the malarial gametes fuse and form the zygote, it then embeds in the wall of this mosquito organ. (7)

22 Cilia beat with this type of wave. (11)

23 Best describes the evolutionary relationships of the different protozoan taxa. (12)

26 Amoebas consume this type of food by using phagocytosis. (5)

27 The process of "sexual" reproduction in ciliates. (11)

30 Conjugation is a unique way of mixing this from two ciliates. (3)

32 Not all amoebas are naked, some live inside these. (5)

34 The pseudopods extend through this opening in the amoeba's test. (6)

36 Complex ciliature involved in locomotion in some polyhymenophora. (5)

38 This stage of the malarial parasite is found in the red blood cells. (11)

39 One of the two ways that the flagella of a mastigophoran beats. (6)

40 When a red blood cell infected with malaria bursts, these are released and they infect other red blood cells. (10)

41 This "cap" is found at the advancing edge of a pseudopod. (7)

42 Another name for a protozoan eyespot. (6)

Down

2 Membranelles of an oligohymenophorans are made up of these. (5)

3 The larger of the two products of gametogony. (3)

5 Single cilia often fuse to form larger structures collectively called this type of ciliature. (8)

6 This type of ciliature around the cytostome is important for identifying the different ciliophoran taxa. (4)

7 The broad, fleshy pseudopods that we usually associate with amoebas are of this type. (8)

8 Describes the cycle of multiple fissions by the malarial parasite inside the human host. (10)

9 The energy source for amoeboid movement. (3)

10 This nucleus is the larger of the two different nuclei found in ciliates. (12)

12 The most common type of reproduction in protozoans. (7)

13 This supports an amoeba's axopod. (7)

14 Process describing the ingestion of rich organic liquid food by protozoans. (11)

15 Microgametes and macrogametes in the malarial life cycle are created by this process. (10)

18 These cilia on a ciliate's body are not involved in feeding and are referred to as this type of ciliature. (7)

21 Undigested food exits from the same point on a ciliate; it's referred to as this. (9)

22 In reticulopod amoebas, their pseudopods form this. (4)

23 The organelle created by the fusion of the lysosome and the ingested food vacuole. (9)

24 This part of the human population is most affected by malaria. (8)

25 This is the more fluid of the two types of protoplasm found in an amoeba. (9)

28 Another name for a contractile vesicle is this type of expulsion vesicle. (5)

29 This type of cell fission creates more than two daughter cells. (8)

31 The number of flagella in a zooflagellate's cell. (3)

33 This element is found in the shells of radiolarians. (6)

35 This is the first type of cell is attacked by the malaria parasite as it enter its human host. (5)

37 As the Reynold's number does this, it gets easier to move through the medium. (5)

38 The number of conjugants involved in conjugation. (3)

PORIFERA

Inside *Digital Zoology*

As you explore the Porifera on the *Digital Zoology* CD-ROM, don't miss these learning tools:

Photos of cross sections, and dissected specimens of preserved sponges *Grantia (Scypha)* and *Leucosolenia* and a variety of other dried sponges show the diverse shapes that sponges have and how their spicules are organized.

Drag-and-drop quiz on the key structures of the sponge *Grantia (Scypha)*.

Defining Differences

Some of the differences described in the following sections are features that arise for the first time in the Porifera, and they define the phylum. Others are things that are important for how animals in the phylum function. Whichever is the case, you'll want to watch for examples of these in the various sponge specimens that you will be examining.

- Cellular grade of organization with
- totipotent cells, including
- choanocytes in an
- aquiferous system.

Cellular grade of organization

That a sponge is an animal may come as a surprise you. You're not alone. Biologists considered them plants until the late 1700s. Sponges are multicellular animals but with a difference. That difference is why they are referred to as **parazoans,** rather than **metazoans**. They are "almost animals" because they are missing one thing that all other metazoans have — cells organized into tissues, and this is why they are described as being functionally organized at the **cellular grade**.

Sponges are an example of one of two ways that animals became multicellular — by using either cells organized into tissues or as aggregates of independent cells. Sponges are an example of the second, and although it's a body plan that serves them well for what they have to do, its dependence on independent cells limits their potential for diversification. It's one of the reasons why you'll see sponges referred to as an evolutionary dead end.

Don't become confused, though, when you are looking at the sponge cross sections in *Digital Zoology*, or your lab. You'll see an outer layer of **pinocoderm** cells and a layer of **choanoderm** cells lining the **spongocoel.** These are separated from each other by the **mesohyl** — but remember these are only layers, not tissues. The reason that they aren't tissue is that cells in a true tissue layer sit on a **basement membrane** and have a variety of cell junctions for intercellular communication — both of these are missing in the sponge cell layers.

Totipotent cells

Even though a sponge's cells are only arranged in layers, there is still a division of labor between them, with certain cells responsible for one function while others have different functions. What all the cells have in common is that sponge cells are **totipotent**, meaning each of the cell types can, if needed, differentiate into the basic **archeocyte** cell and then redifferentiate into a whole new type of cell. It's part of how sponges respond to changes in their environment and the reason for their often asymmetric shapes. They can't move, so they grow toward where they want to be. If they're damaged they can grow back what was lost. This is an effective strategy, but its dependence on totipotency also prevents the development of tissues, a characteristic of all the other multicellular body plans that follow.

✍ *What are the different types of cells that you would find in a sponge, and what does each cell type do?*

Choanocytes

Pumping water and filtering food from it is what sponges do best, and it's their **choanocytes** that take care of that. These specialized cells have a central **flagellum** surrounded by a **collar** of **microvilli** connected to each other by an even finer mesh of **microtubules**. As the central flagella beats, water is drawn in toward the collar and between the microvilli and its **microtubular mesh**. Only water can pass through, and small particles of food carried in it become trapped on this delicate sieve. The choanocytes then use **phagocytosis** to consume the trapped food.

It seems so simple, but you multiply this by the huge numbers of choanocytes in a sponge and you have lots of water being pumped and just as much food being filtered from the water. Sponges were some of the first filter feeders in the ancient oceans, and the reason sponges have survived is that, in many ways, they still are the best at filter feeding.

✍ *How does captured food get from a choanocyte to the other sponge cells?*

Aquiferous system

There's more to it than pumping and filtering water, however. The nutrients need to be extracted from the water. That's the role of the **aquiferous system** which combines pumping and food capturing choanocytes with a variety of canals, chambers, and spaces through which water flows. In sponges the aquiferous system is arranged into three different architectures: **asconoid**, **syconoid,** and **leuconoid**.

In the simplest, the asconoid, choanocytes line the **spongocoel,** and water is drawn in through cells in the body wall. There's always a possibility that some water could make it to the central spongocoel without being filtered. In the syconoid architecture, the filtering choanocytes are found in smaller radial canals that empty into the central spongocoel, which is a better way to capture particulate food.

But as the diameter of the passages through which the water flows narrows, and the same volume is pumped, the water speeds up. It's like when you put your finger over the end of the garden hose to squirt the water farther. It works because the water flows faster through the narrower opening at the end of the hose. Syconoid sponges are better at trapping and pumping than the asconoid architecture but, there's still room for improvement.

That improvement is the leuconoid sponge. Here the narrow tubes through which the water flows widen into chambers where the choanocytes are located. Water rushes through the sponge and slows in the chambers where it is filtered. It leaves the chamber and travels through more of these choanocyte spaces before leaving the sponge through the multiple oscula. Now you can see why most of the sponges have the leuconoid architecture and why we said earlier that sponges are pretty good at what they do.

 *Use the following terms and place them in the proper order to explain the flow of water through the three different sponge architectures: spongocoel, osculum, radial canal, excurrent canal, prosopyle, apopyle, choanocyte chambers, porocyte.*

A Closer Look Inside *Digital Zoology* - Porifera

Leucosolenia

Don't let the name fool you. *Leucosolenia* is not a leuconoid sponge! It's an example of the simplest of sponge architectures. The vase shape may be hard to see in most preserved specimens because the species is colonial. The central spongocoel of each sponge is connected to the others by hollow horizontal branches, extensions of the spongocoel. Either with the specimens you have in the lab or the photos in *Digital Zoology* take a look at the specimen, and you'll see that it's made of delicate spicules over which is woven a fine film of living tissue, making it difficult to identify the different cell types.

Grantia *or* Scypha

Either *Grantia* or *Scypha* are excellent examples of this intermediate architecture and a chance to see the cell layers of a sponge. What most students find hard when observing this specimen is that it is a modified version of the syconoid architecture. Look closely at the surface of a preserved specimen and you'll see that it's covered in fingerlike projections. The hollow inside of each finger is the radial canal and the spaces between the fingers are the incurrent canal. In the dissected specimen in *Digital Zoology* you can see the inner surface of the spongocoel and the openings of the radial canals contained in the fingers that lead to the spongocoel.

Another problem in understanding how this specimen works is apparent when looking at the cross section. As microscopic sections are made, the cut isn't perfect, leaving a nice petal-like array of radial canals surrounding the central spongocoel. Instead, you'll have to look around on your specimen to see the connection between the spongocoel and the radial canal. It is always a good idea to look at more than one section whenever possible. Once you can identify the inside of the radial canal, you'll see the thicker choanocytes on the inner surface. If you have good samples and an adequate microscope, the flagella may be visible, but don't expect them to stick out like they do in the drawings. Instead they usually appear as a tangled mat, at or near the cell body of the choanocyte.

A second problem with the sectioned materials is that instead of cutting through the spicules, the blade often breaks them, and this tears the adjacent tissue. Move along the inner surface of the radial canal to find the connection between it and the incurrent canals. Be patient; it's there!

Other sponges, spicules, and gemmules

Almost all living sponges have this leuconoid architecture, but it's difficult, especially with dried specimens, to see all the elements of their design. Larger canals that lead to the osculum are much easier to identify than the flagellated chambers.

Sponges support themselves with spicules, and the different classes in the phylum are defined by their spicules. You'll have seen broken spicules in *Grantia* or *Schypha* cross sections. Take a closer look at the surface of some of the other sponges or microscope slides of prepared spicules to see some of the different shapes. Perhaps the most elegant spicules are the six-sided spicules that glass sponges weave into their elaborate skeletons. You'll find some good examples on the *Digital Zoology* CD.

✍ *Why do freshwater sponges have gemmules as part of their life cycle?*

Cross-References to Other McGraw-Hill Zoology Titles

Integrated Principles of Zoology, 11th edition. C.P. Hickman, L.S. Roberts & A. Larson. Chapter 12.

Animal Diversity, 2nd edition. C.P. Hickman, L.S. Roberts & A. Larson. Chapter 5.

Zoology, 5th edition. S.A. Miller & J.P. Harley. Chapter 9.

Biology of the Invertebrates, 4th edition. J. Pechenik. Chapter 4.

Laboratory Studies in Integrated Principles of Zoology, 10th edition. C.P. Hickman, F. Hickman & L. Kats. Chapter 7.

General Zoology Laboratory Guide, 13th edition. C. Lytle. Chapter 6.

Structures Checklist

Here are some of the structures that you should be able to easily find in *Digital Zoology* and the specimens that you will be looking at in your lab. After reading your lab handout, you might want to add more and, depending on the equipment available in your lab, you might see more. As you study the material, you might also want to make some notes on how some of these structures looked or include a drawing in your lab notes. (Structures indicated by * may be hard to see.)

Asconoid sponge: Leucosolenia

Preserved specimen	Cross section	Additional structures
☐ Body wall with spicules	☐ Body wall with spicules	☐
☐ Osculum	☐ Choanocyte surface	☐
☐ Spongocoel	☐ Porocytes*	☐
☐	☐ Spongocoel	☐
☐	☐	☐

Syconoid sponge: **Grantia** *or* **Scypha**

Preserved specimen

- ☐ Apopyles
- ☐ Body wall
- ☐ Incurrent canals
- ☐ Osculum
- ☐ Radial canals
- ☐ Spicules
- ☐ Spongocoel
- ☐
- ☐
- ☐

Cross section

- ☐ Amphiblastula larvae*
- ☐ Apopyles
- ☐ Choanocytes
- ☐ Choanocyte flagella*
- ☐ Incurrent canals
- ☐ Mesohyl
- ☐ Prosopyles
- ☐ Radial canals
- ☐ Spicules
- ☐ Spongocoel

- ☐
- ☐
- ☐

Additional structures

- ☐
- ☐
- ☐
- ☐
- ☐
- ☐
- ☐

Sponges, spicules and gemmules

- ☐ Monoaxon
- ☐ Triaxon
- ☐ Hexaxon
- ☐ Triradiates
- ☐ Polyaxon
- ☐ Spongin
- ☐ Gemmule
- ☐
- ☐

Crossword Puzzle – Porifera

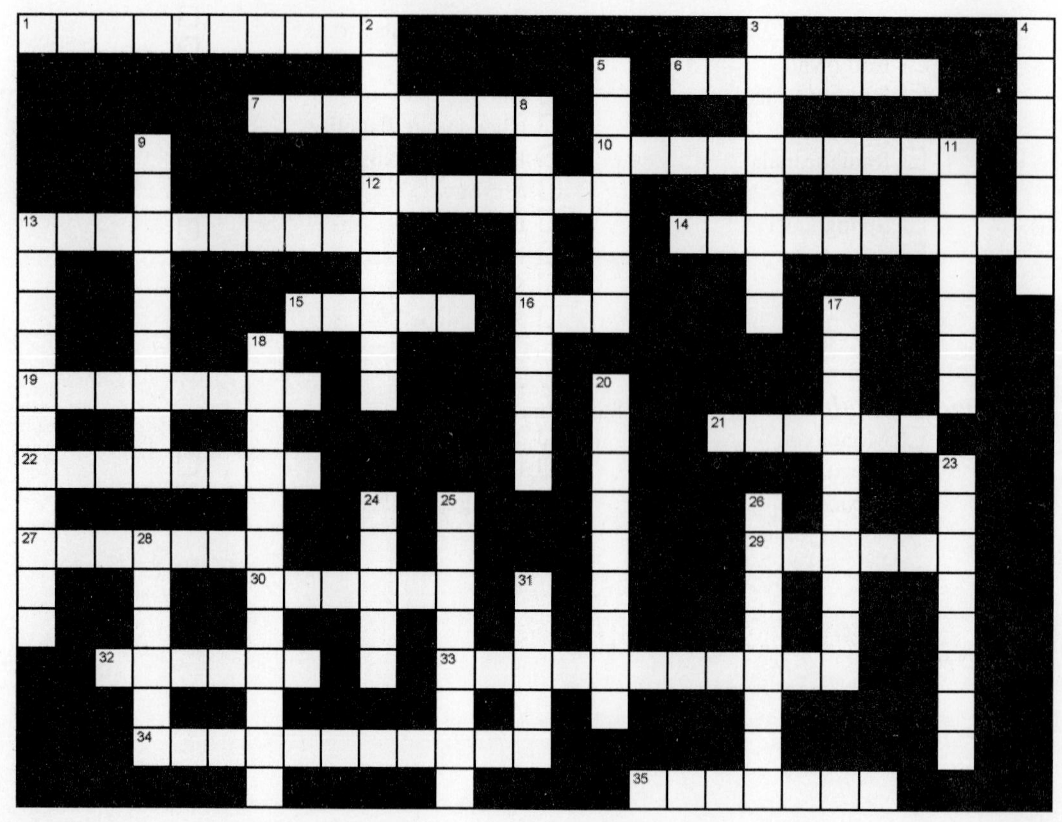

Puzzle solution. An interactive web based version of this puzzle, and its solution, are available on the *Digital Zoology* web site at www.mhhe.com/DigitalZoology/Students. With the interactive puzzle you can check to see if individual words or the whole puzzle is correct, and get hints for single letters.

Across

1 Although it's not a tissue, because sponges don't have them, the cells lining the spongocoel are collectively called this. (10)

6 Found between the inner and the outer layers of cells in a sponge. (7)

7 The supporting and defensive structures in sponges. (8)

10 These large cells found in the larval stage of some sponges will form the outer choanoderm. (10)

12 The opening, in all types of sponge architectures, that water pumped by the sponge exits. (7)

13 The name for the outer layer of cells in a sponge. (10)

14 This cavity is lined with choanocytes in asconoid sponges but they're missing in the same cavity in the syconoid form. (10)

15 The number of different sponge architectures. (5)

16 Unlike a leuconoid sponges, asconoid and syconoid sponges have how many oscula? (3)

19 Sponges are organized at the this grade. (8)

21 Sponges are becoming of interest because not only do they protect themselves with spicules but also produce these to defend themselves. (6)

22 In leuconoid sponges, the choanocytes are found inside these. (8)

27 Sponges are organized at the cellular grade and don't have cells organized as these. (7)

29 An amphiblastula is this stage in the life cycle of some sponges. (6)

30 The microvilli of these cells form this part of the choanocyte. (6)

32 Sponges are called this type of feeder because they do this to the water that they pump. (6)

33 The unique cells in a sponge that propel water through it. (11)

34 Specialized types of these cells transport sperm caught by the collar cell to the egg in a sponge. (11)

35 Describes multicellular animals that lack tissues. (7)

Down

2 These flagellated cells in the embryo of some sponges will form the choanoderm. (10)

3 Sponge architecture consisting of choanocytes in the spongocoel and a single osculum. (8)

4 Water flows out this opening into the spongocoel of a syconoid sponge. (7)

5 This dormant sponge structure helps a freshwater sponge get through the winter. (7)

8 The sponge cell that produces the spicules. (10)

9 The structure on a choanocyte that propels the water through a sponge. (9)

11 The protein constituent of sponge spicules. (7)

13 These cells form the outer surface of a sponge. (11)

17 Another name for the system of canals and chambers inside a sponge. (10)

18 Sponge spicules can be grouped into two types, the small microscleres that reinforce or pack of the sponge body and these larger spicules. (12)

20 One of the two types of canals in a syconoid sponge. This one is not lined with choanocytes. (9)

23 Compared to asconoid and syconoid sponges, the leuconoid sponges have ______ oscula (osculum). (8)

24 Compared to the hollow coeloblastula, the parenchymula larva in a sponge is best described as this. (5)

25 Water enters an asconoid sponge through this special sponge cell. (8)

26 This hollow larval stage of some sponges resembles the developmental stage in higher animals. (8)

28 Sponges found in the deepest parts of the oceans have spicules made of this. (6)

31 As water moves through the choanocyte chambers it does this so that it can be filtered. (5)

Self Test - Porifera

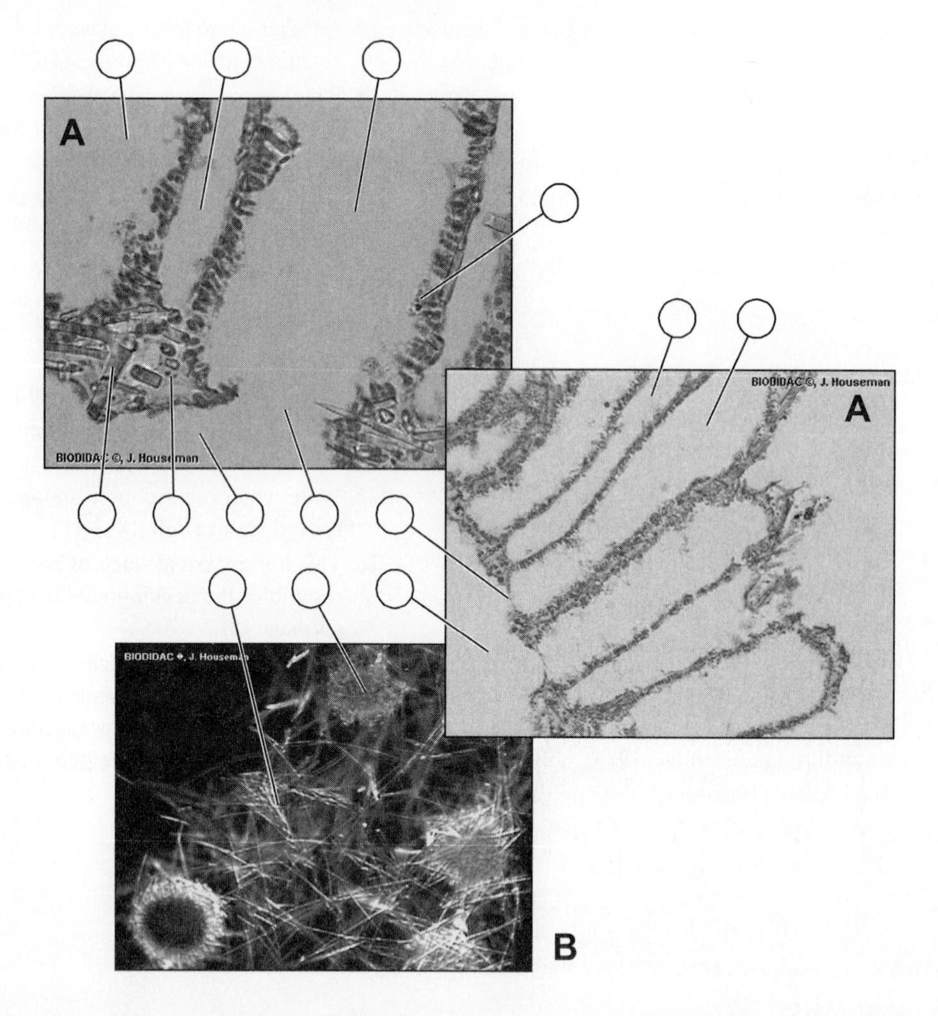

Use the following labels to identify the photographs. You may have to use a label more than once, and some labels may not be appropriate for the photographs. Answers are available on the *Digital Zoology* web site at www.mhhe.com/digitalzoology. Be sure to try the interactive Drag-and-Drop quizzes that are available on the *Digital Zoology* CD-ROM. A color version of this Self Test is available in the Adobe Acrobat version of the Student Workbook in the workbook folder on the *Digital Zoology* CD-ROM.

Specimens

A- *Grantia (Scypha)*

B- *Spongilla*

Labels

1) Apopyle
2) Choanocyte chamber
3) Choanocytes
4) Gemmule
5) Incurrent canal
6) Mesohyl
7) Osculum
8) Porocyte
9) Prosopyle
10) Radial canal
11) Spicule
12) Spongocoel

CNIDARIA

Inside *Digital Zoology*

As you explore the Porifera on the *Digital Zoology* CD-ROM, don't miss these learning tools:

Photos of preserved specimens and cross sections of hydrozoans including: *Hydra*; *Obelia*; *Gonionemus*; *Pennaria;* and the Portuguese Man-of-War, *Physalia.* The jellyfish *Aurelia*, represents the Schyphozoa and the sea anemone, *Metridium* the anthozoan body plan. Take a look at a variety of different corals and explore the various stages in the cnidarian life cycle with the interactive life cycles.

Video of *Hydra* and how it moves and changes shape.

Drag-and-drop quizzes on the key structures of *Hydra, Aurelia,* and *Metridium.*

Interactive cladogram showing the major events that gave rise to the four cnidarian classes. A summary of the key characteristics of each class are combined with an interactive glossary of terms.

Defining Differences

Cnidarians are an important group for us to study because their unique features are intermediate between the first multicellular organisms that preceded them, the Porifera, and those that followed. Some of the features in the following list appear for the first time in the Cnidaria, and they define the phylum. Other characters in the list are things that we see for the first time in animals and are important for understanding how organisms in the phylum function. Whichever is the case, you'll want to watch for examples of these in the various cnidarian specimens that you will be examining.

- Radially symmetric animals organized at the
- diploblastic tissue-grade organization with unique
- cnidocytes, an
- incomplete gut that can be used as a
- hydrostatic skeleton, and a
- dimorphic life cycle.

Radially symmetric

Radially symmetric animals do not consistently orient in any particular direction and don't have an anterior or posterior end like **bilaterally symmetric** animals do. Instead, cnidarians have oral and aboral surfaces. One of the most distinctive features of the Cnidaria is their radial symmetry, and it gives us some clues to how they function.

One consequence of the absence of directed movement is that these animals have a nerve net distributed throughout the body rather than being complex and centralized at the anterior end of the animal, which is the case in bilaterally symmetric animals. If you don't always face or move in the same direction, there's no advantage to concentrating sensory structures to detect what is ahead. Instead, sensory structures are evenly spaced around the whole organism allowing it to react equally well in all directions. Watch for the position of **rhopalia,** tentacles, and any other structures radially arranged in the specimens you look at in your lab and in *Digital Zoology*.

One group of cnidarians take radial symmetry a step farther. The anemones and corals (Anthozoa) show **biradial symmetry**. Even though most of their structures are radially arranged around the oral-aboral axis, there is at least one important paired structure that restricts the planes of symmetry to only two. When you look at the anemone watch for the **syphonoglyphs** and the arrangement of the internal **septa** to see why these animals are biradial, rather than radial.

One of the assumptions often made about **primitive** and **advanced** traits in the evolution of animals is that radial symmetry is primitive and bilateral is advanced. It's based on the idea that radially symmetric animals are **pelagic**, passively floating in their marine environments collecting food as they bump into it, or sessile, sitting on the bottom waiting for food to float by. Pelagic animals became sessile, then started to move across the substrate. The anthozoan's shift to biradial symmetry could be an intermediate step on the way to the more advanced bilateral symmetry.

Not everyone agrees, and an alternate would see the first animals being motile and bilaterally symmetric with some adopting a sessile existence and the benefits of radial symmetry for that type of life. It's not all that far fetched. We will see later that part of the sequence is exactly what the echinoderms did.

 *What other types of symmetry can animals have?*

Diploblastic tissue-grade organization

The cnidarians are the flip side to the multicellular coin, being the first animals to have tissues derived from **endoderm** and **ectoderm**. They only have two of the possible three germ layers, and that's why we refer to them as **diploblastic**. The next change we'll see is the presence of the third germ layer, the **mesoderm**, and being **triploblastic** with all three cell layers.

These animals are organized at the **tissue grade** because neither ectoderm nor endoderm have been modified to form organ systems. That doesn't mean their tissues are simple, with all cells doing the same thing. There is division of labor between the cells. A good example of this can be seen in the various cells found in the outer **epidermis** and the inner **gastrodermis**.

 *What's the difference between a mesoglea and a mesenchyme found between the two layers of a diploblast?*

The outer surface of the animal is defined by the **epitheliomuscular cells** that connect with each other to form the body covering. As you might have guessed from their name, these cells do more than that. They also have contractile **myonemes** at the base of the cell, and these are aligned in the same direction. Nerve cells, also formed from the ectoderm, wind between the epitheliomuscular cells to create the nerve net.

Endoderm cells form the **gastrodermis** that lines the **gastrovascular cavity (coelenteron)**, and they, too, have a variety of functions. For example, **nutritive muscular cells** are responsible for phagocytic uptake of food from the gastrovascular cavity. Their cell bases also have myonemes arranged in the opposite direction to those of the epitheliomuscular layers. Another type of cell found in the gastrodermis are gland cells, which secrete digestive enzymes into the gastrovascular cavity to start the digestion of ingested food.

Cnidocytes

The most specialized cell found in a cnidarian gives the phylum its name. These are the stinging **cnidocytes** embedded in the epidermis around the mouth and, when present, the tentacles, or arms. Cnidarians are predators and capture their prey using these unique stinging cells. The secret to how they work is the cell's specialized organelle, the **nematocyst**, found near the cell surface. Coiled inside the cnidocyte's nematocyst is a deadly barbed, sticky, or poison-tipped, thread waiting to be fired if the cell is triggered. When fired, they shoot out at tremendous velocities easily penetrating, or entangling, potential prey. Once captured, prey is put inside the gastrovascular cavity where it is digested.

Incomplete gut

The digestive tract of a cnidarian is an **incomplete gut** because it has only one of the two possible openings that guts have — a mouth, but no anus. Once food is in the gut, the mouth closes, and the cells of the gastrodermis begin to work. Digestive enzymes are secreted into the gastrovascular cavity. The first part of the digestive process breaks down the food into particles small enough for the nutritive muscular cells to consume by phagocytosis — the final stage of the digestive process.

That's the simple view of the digestive system. A lot of cnidarians are colonial, and each member of the colony is connected to the next by extensions of the gastrovascular cavity. The flagellated cells of the gut ensure that food is distributed to everyone in the colony using these common connections. Jellyfish have a different set up, and their gut isn't as much a cavity as a series of canals that allow the nutrients to flow and reach all the different parts of the animal.

There is at least one disadvantages to this type of gut — any undigested food must be emptied from the gut through the mouth.

Hydrostatic skeleton

Cnidarians use their gut for more than digestion, and we see this new functional system for the first time in the phylum, a skeleton — a **hydrostatic skeleton**. When a cnidarian's mouth is closed, the gastrovascular cavity is a closed fluid-filled space, and if muscles surrounding the cavity contract, they'll change the shape of the space.

Think of it as a water-filled cylindrical balloon with two layers of contractile cells, one with the myonemes running the length (longitudinal) of the balloon and the other arranged in circles (circular) around the balloon's diameter. When the longitudinal ones contract, they shorten and the diameter of the balloon increases. The opposite happens when the circular myonemes contract. The balloon becomes a longer and narrower cyclinder. But there is more than just a change in shape going on, and it's why we call this a skeleton.

Myonemes, or muscles, burn energy to contract and shorten, but they have no way to return to their original length unless something stretches them. That's the role of the skeleton with its two sets of antagonistic muscles. When one contracts it causes the other to lengthen, and in a hydrostatic skeleton the muscles use fluid to do that. That increasing diameter that comes from contraction of the longitudinal myonemes stretches the circular muscles back to their original length. The result of this is that these changes in shape allow for movement.

Dimorphic life cycle

As it changes from a sessile polyp to the mobile medusa, a cnidarian undergoes a tremendous transformation in its appearance and how it functions. It's why their life cycle is **dimorphic,** and for the first cnidarians, both stages were equally important to how the animal lived. As the group evolved, the importance of the two stages changed. In some cnidarians, the animal's entire life is spent as a polyp; in others, as a medusa; and in still others, something in between.

✍ *What are some of the different roles the medusa and polyp have in the different cnidarian classes?*

The larval stage of the cnidarians is a planula, a ciliated solid ball of cells that forms from the fertilized egg. The planula is motile, and after swimming through the water for a while it settles down on the substrate and develops into the polyp. The cnidarian planula is often used in theories of the origins of the metazoa, so be sure to take a look.

✍ *Why is the planula important in theories of metazoan evolution?*

A Closer Look Inside *Digital Zoology* - Cnidaria

Hydrozoa: Obelia *and* Hydra

First appearances can be deceiving. A quick look at the preserved specimen of *Obelia* and you might think you are looking at some sort of plant. (If you don't have a preserved specimen, take a look inside *Digital Zoology* for an example.) If you have a preserved specimen, take a closer look under the microscope and you'll see the different **zooids** (hydroids) that cover the stalks. Even with the preserved specimen you may find it hard to see any details of the zooids because they are covered with a nonliving protective **perisarc**. In prepared slides, the living tissue, **coenosarc**, is stained and you can see it through the perisarc of the animal that makes up the colony.

Almost all hydrozoans form some sort of colony, and polyps in the colony often have different jobs to do — another example of division of labor. The two most common jobs for polyps are feeding and reproduction. The **gastrozooids** have tentacles and feed, and the **gonozooids** produce the small medusa that will grow into the reproductive free-swimming medusa. Each of the polyps is connected to the other through the stems and branches of the colony.

Hydra is often used in the zoology labs to show the various features of a cnidarian polyp, and it's a great specimen for doing that. What may be confusing is that this hydrozoan's life cycle is spent entirely in the polyp stage; there's no medusa. So why is it a hydrozoan? It's important to remember that ancestral characteristics of a group can be modified in a variety of ways, and the original character often doesn't look at all like the trait when it first appeared. The most drastic modification, a trait's complete disappearance, is an example of an advanced condition. An absence of the medusa doesn't disqualify *Hydra* from being a member of the Hydrozoa. With the reproductive medusa gone, the polyp has reproductive responsibilities, and you'll see examples of both asexual and sexual reproduction in *Digital Zoology* and your lab materials.

If they aren't available in your lab, *Digital Zoology* has a variety of other colonial hydrozoans at which you can take a look. One of these is *Gonionemus,* where you'll see the velum, also a hydrozoan trait.

✍ *Why is the velum important in how the hydrozoan medusa swims?*

The most spectacular colonial hydrozoan is the Portuguese man-of-war, *Physalia.* These large hydrozoans drift in the oceans using a modified polyp filled with air that acts as both a float and sail. Inflated with gas, it keeps the colony at the water surface, and as the winds blow against the part that's above the water line, the same modified polyp acts a sail. Dangling behind are a variety of tentacles armed with cnidocytes. Clustered among these, are the feeding and reproductive polyps of the colony.

✍ *How many different types of polyps does the Portuguese man-of-war have?*

Scyphozoa: The jellyfish Aurelia

The jellyfish life cycle, like all scyphozoans, has the usual cnidarian stages. The difference is that in this class the polyp is highly modified and only around for a short time when compared to the longevity of the medusa.

Aurelia's gonads are located next to the four gastric pouches, and gametes are released into the gastrovascular cavity and out the mouth. The planula swims around for a while before it settles down on the substrate, and changes into the unusual polyp stage, the **scyphostome**. This grows and then undergoes transverse fission creating the **strobila,** which look a stack of dinner plates piled one on top of the other. Each of the "plates" develops small arms and then separates from the top of the strobila as an **ephyra,** which ultimately grows into the mature adult jellyfish whose shape with which you might be more familiar.

As jellyfish float through the water, small zooplankton are trapped in mucuos on the underside of the bell where cilia, along with the help of the oral arms, move the food into the mouth and the gastrovascular cavity. Compared to a polyp, the gastrovascular cavity is extensively modified, and the mouth leads to four gastric pouches connected to a complex series of canals that lead to the radial canal around the margin of the bell. These canals work like a circulatory system, and food follows a well-defined path as it moves through the different parts of the bell of the medusa and distributes nutrients to the tissues.

What moves the fluid in a jellyfish's gastrovascular cavity?

Anthozoa: The anemone Metridium

If you've been working your way through the invertebrates phylogenetically in either your lab or *Digital Zoology*, you're probably glad to finally get to look at a decent-sized animal like the sea anemone. But remember, this is one of the two different types of anthozoa, and the small-size delicate corals — also anthozoans — have a huge impact on the marine environment building of coral reefs, some of the most productive marine environments. The most striking thing about anemones is their size, especially for a diploblastic organism. One of things you'll want watch for as you work with the specimen are adaptations that allow these animals to be this big.

The mouth is surrounded by tentacles that trap food, and then push it into the gastrovascular cavity where it is digested. Take a close look at the mouth and you'll see that it isn't a perfect circle; it's more oval in shape. At the edges of the mouth are paired ciliated grooves, the **syphonoglyphs,** which move fresh oxygenated water into the gastrovascular cavity, and that's important for the complex of tissues found inside the cavity. The presence of these functionally important paired structures creates a new type of symmetry that we haven't seen before, biradial symmetry.

Internally, the gastrovascular cavity is divided by either **incomplete septa** or **complete septa** that increase the surface area available for digestion of food. It's the incomplete septa that are important for digestion. The free ends of the septa have three lobes where there are glandular cells that release the digestive enzymes, ciliated cells to mix the content of the cavity, and cnidocytes. Anemones consume large prey, and having cnidocytes inside the digestive tract is a big help in killing prey that may still be alive when ingested. Anthozoans are the only cnidarians with cnidocytes in the gastrovascular cavity.

Complete septa do a little more than connect with the pharynx. They also have retractor muscles that help change the shape of the anemone, mix the contents of the gastrovascular cavity, and help to move the tentacles that surround the mouth. If you have trouble telling the difference between these different parts of the cavity in your dissected specimen, take a look at the cross section slide if it's available in your lab, or take a look in *Digital Zoology*.

If cnidarians are diploblastic, lacking mesoderm, how can an anemone have retractor muscles?

Cross-References to Other McGraw-Hill Zoology Titles

Integrated Principles of Zoology, 11th edition. C.P. Hickman, L.S. Roberts & A. Larson. Chapter 13.

Animal Diversity, 2nd edition. C.P. Hickman, L.S. Roberts & A. Larson. Chapter 6.

Zoology, 5th edition. S.A. Miller & J.P. Harley. Chapter 9.

Biology of the Invertebrates, 4th edition. J. Pechenik. Chapter 6.

Laboratory Studies in Integrated Principles of Zoology, 10th edition. C.P. Hickman, F. Hickman & L. Kats. Chapter 8.

General Zoology Laboratory Guide, 13th edition. C. Lytle. Chapter 7.

Structures Checklist

Here are some of the structures that you should be able to easily find in *Digital Zoology* and the specimens that you will be looking at in your lab. After reading your lab handout, you might want to add more and, depending on the equipment available in your lab, you might see more. As you study the material, you might also want to make some notes on how some of these structures looked or include a drawing in your lab notes. (Structures indicated by * may be hard to see.)

Obelia

Hydroid colony

- ☐ Blastostyle
- ☐ Coanosarc
- ☐ Gastrovascular cavity
- ☐ Gastrozooid (hydranth)
- ☐ Gonotheca
- ☐ Gonozooid (gonangia)
- ☐ Hydrocalus
- ☐ Hydrorhiza
- ☐ Hydrotheca
- ☐ Hypostome
- ☐ Medusa buds
- ☐ Mouth
- ☐ Perisarc
- ☐ Tentacles
- ☐
- ☐
- ☐

Medusa

- ☐ Manubrium
- ☐ Medusa
- ☐ Nematocysts
- ☐ Radial canals
- ☐ Statocysts
- ☐ Tentacles
- ☐
- ☐

Additional structures

- ☐ Planula
- ☐
- ☐

Hydra

- ☐ Basal disc
- ☐ Buds
- ☐ Cnidocytes on tentacles
- ☐ Epidermis
- ☐ Epitheliomuscular cells*
- ☐ Gastrodermis
- ☐ Gastrovascular cavity (Coelenteron)
- ☐ Gland cells*
- ☐ Hypostome
- ☐ Interstitial cells*
- ☐ Mesoglea
- ☐ Mouth
- ☐ Nutritive-muscular cells*
- ☐ Ovaries
- ☐ Spermaries (testes)
- ☐ Tentacles
- ☐
- ☐

Additional structures

- ☐ Planula
- ☐
- ☐

Aurelia

Adult

- ☐ Gastric pouches
- ☐ Gonads
- ☐ Lappet
- ☐ Marginal tentacles
- ☐ Oral arms
- ☐ Oral tentacles
- ☐ Radial canal
- ☐ Rhopalia
- ☐ Ring (circular) canals
- ☐ Stomach
- ☐
- ☐
- ☐
- ☐

Life cycle

- ☐ Planula
- ☐ Schyphistome
- ☐ Ephyra
- ☐ Strobila
- ☐
- ☐
- ☐

Additional structures

- ☐
- ☐
- ☐

Gonionemus

- ☐ Cnidocytes
- ☐ Gonad
- ☐ Manubrium
- ☐ Mouth
- ☐ Radial (circular) canal
- ☐ Radial canal
- ☐ Ring canal
- ☐ Tentacles
- ☐ Velum
- ☐
- ☐

Additional structures

- ☐
- ☐

Metridium

Cross section

- ☐ Ciliated lateral lobe of septa
- ☐ Cnidoglandular lobe of septa
- ☐ Complete septa
- ☐ Epidermis
- ☐ Gastrodermis
- ☐ Gastrovascular cavity (Coelenteron)
- ☐ Incomplete septa
- ☐ Mesenchyme (Mesoglea)
- ☐ Pharynx
- ☐ Retractor muscles
- ☐ Siphonoglyphs
- ☐
- ☐
- ☐

Preserved specimen: Intact and dissected

- ☐ Acontia filaments
- ☐ Complete septa
- ☐ Gastrovascular cavity (Coelenteron)
- ☐ Incomplete septa
- ☐ Mouth
- ☐ Oral disc
- ☐ Pedal disc
- ☐ Pharynx
- ☐ Retractor muscle
- ☐ Siphonogyphs
- ☐ Tentacles
- ☐
- ☐
- ☐

Additional structures

- ☐
- ☐
- ☐
- ☐
- ☐

Crossword Puzzle - Cnidaria

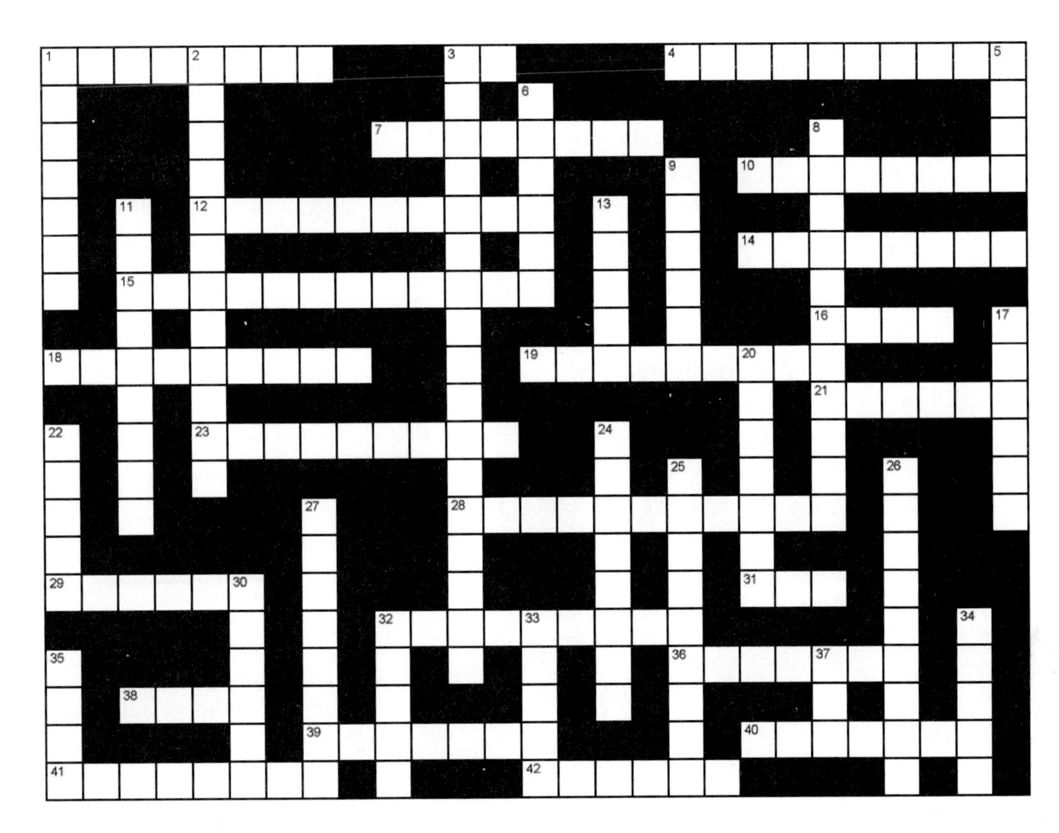

Puzzle solution. An interactive web based version of this puzzle, and its solution, are available on the *Digital Zoology* web site at www.mhhe.com/DigitalZoology/Students. With the interactive puzzle you can check to see if individual words or the whole puzzle is correct, and get hints for single letters.

Across

1 The outer nonliving covering of a colonial hydrozoan. (8)

3 Is the perisarc in a colonial cnidarian living material? (2)

4 In thecate hydrozoans the gastrozooid is surrounded by this nonliving structure. (10)

7 Unlike all the other cnidarian classes, in this one cnidocytes are found inside the gastrovascular cavity. (8)

10 Many hydrozoans form these with division of labor between the different polyps. (8)

12 Colonial hydrozoans resemble plants, and the "roots" that connect the different parts of the colony are called this. (10)

14 This embryonic tissue layer is missing in a diploblast. (8)

15 These polyps on a colonial hydrozoan are responsible for feeding. (12)

16 If you are looking straight into the mouth of a hydrozoan, this surface is facing you. (4)

18 This nonliving structure surrounds the gonozooid of a thecate hydrozoan. (9)

19 Gastrozooids are also called this. (9)

21 Cnidarians bodies are organized at this grade; they have these but still don't have organ systems. (6)

23 The oral opening is located at the tip of this tubular extension in a scyphozoan medusa. (9)

28 The cnidocyte squeezes to fire the nematocyst—the ______ hypothesis for nematocyst firing. (11)

29 Although it is a hydrozoan, *Hydra* is missing this stage of the typical hydrozoan life cycle.

It's thought to be an adaptation to living in fresh water. (6)

31 Describes the nervous system in a cnidarian. (3)

32 A statocyst always contains one of these. (9)

36 To fire the nematocyst, the permiability of the cnidocyte cell changes in some way. This is a process referred to as this hypothesis for cnidocyte firing. (7)

38 Cubozoans have quadraradial symmetry, meaning their symmetry is based on this number. (4)

39 Epitheliomuscular cells are found on this side of the cnidarian body wall. (7)

40 Located at the base of an anemones (Anthozoa) gastrovascular cavity, these help kill the ingested prey. (7)

41 The layer, remember it's not a tissue, found between the inner and outer tissue layers of a cnidarian. (8)

42 The type of symmetry characteristic of cnidarians. (6)

Down

1 This ciliated solid ball, a cnidarian larva, may be a clue to the origins of multicellular animals. (7)

2 In the scyphozoan life cycle, once the planula settles it turns into this. (12)

3 The dual roles of cells lining the cnidarian coelenteron is reflected in their name. (17)

5 A cnidarian gastrovascular cavity is an incomplete gut because it lacks this. (4)

6 These delicate minute anthozoans create a unique marine environment. (6)

8 Small medusa form on this central rod found in some gonozooids. (11)

9 This free-swimming stage in the schyphozoan life cycle matures into an adult medusa. (6)

11 The nematocyst is this type of subcellular structure. (9)

13 These cells secrete digestive enzymes into the digestive cavity of a cnidarian. (5)

17 A "true nematocyst" injects proteins and this toxic compound when it penetrates its prey. (6)

20 The energy to fire the nematocyst forms as the organelle develops — the _____ hypothesis for cnidocyte firing. (7)

22 As the medusa contracts, this structure decreases the opening through which water is expelled. It's also a characteristic of a hydrozoan medusas. (5)

24 Horizontal divisions of this stage of the schyphozoan life cycle make it resemble a stack of plates. (8)

25 The medusa dominates in this cnidarian's life cycle. (9)

26 This unique cell gives the Cnidaria its name. (9)

27 In this cnidarian class, the medusa and polyp are equally important in the life cycle. (8)

30 Instead of dorsal and ventral sides, cnidarians have oral and this surface. (6)

32 The gastrovascular cavity of an anthozoans is divided by these. (5)

33 The lobe in anthozoan incomplete septa where you would find the cnidocytes. (5)

34 Corals and other anthozoans sit on this disk. (5)

35 This part of a plant is most like the hydrocalus of a hydrozoan colony. (4)

37 Biradially symmetric anthozoans have this many syphonoglyphs; it's one of the reasons they have biradial rather than radial symmetry. (3)

Self Test - Cnidaria

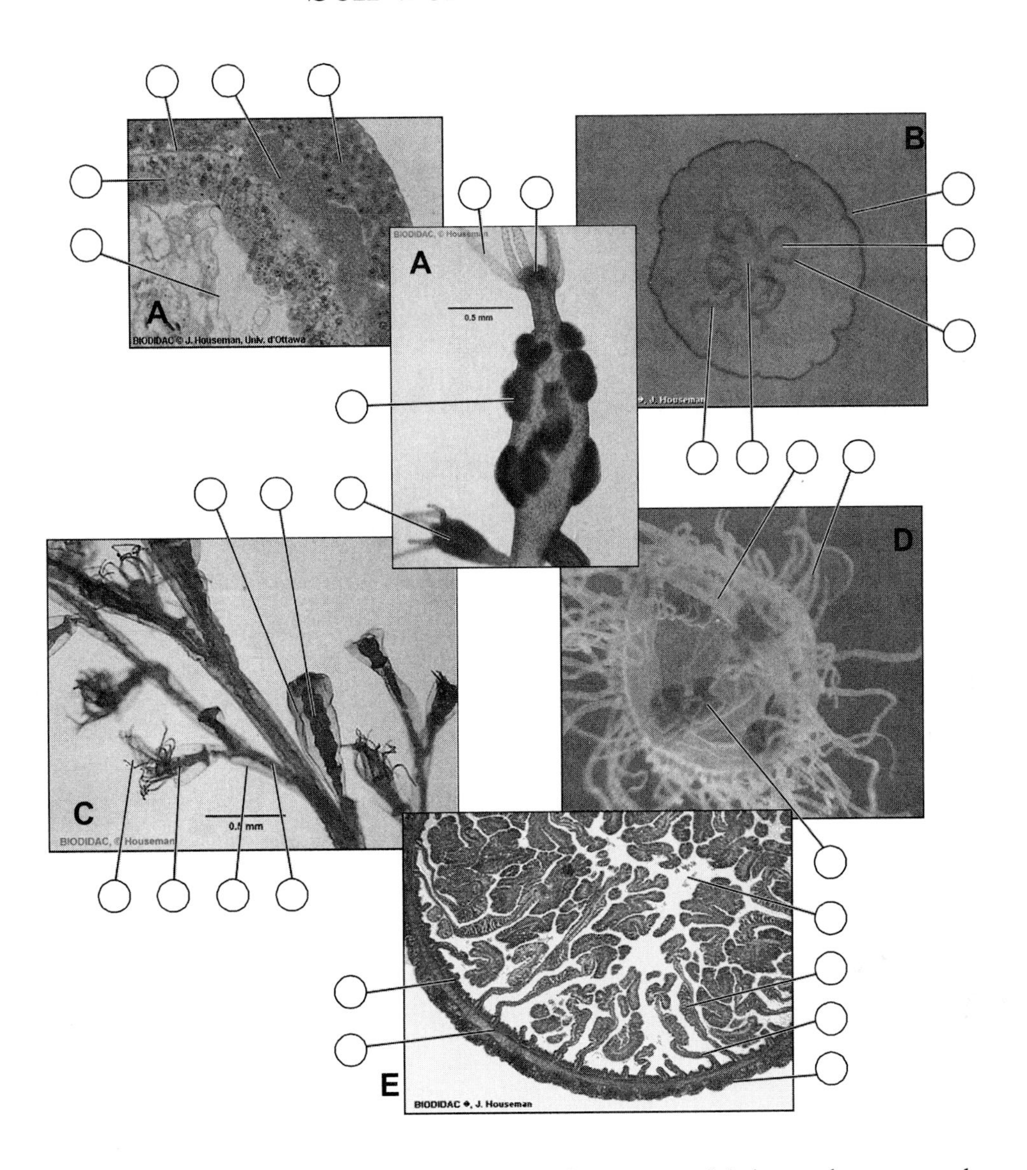

Use the following labels to identify the photographs. You may have to use a label more than once, and some labels may not be appropriate for the photographs. Answers are available on the *Digital Zoology* web site at www.mhhe.com/digitalzoology. Be sure to try the interactive Drag-and-Drop quizzes that are available on the *Digital Zoology* CD-ROM. A color version of this Self Test is available in the Adobe Acrobat version of the Student Workbook in the workbook folder on the *Digital Zoology* CD-ROM.

Specimens

A- *Hydra*
B- *Aurelia*
C- *Obelia*
D- *Gonionemus*
E- *Metridium*

Labels

1) Bud
2) Coenosacrc
3) Epidermis
4) Gastric pouch
5) Gastrodermis
6) Gastrovascular cavity
7) Gastrozooid
8) Gonad
9) Gonotheca
10) Gonozooid
11) Hypostome
12) Incomplete septa
13) Mesoglea
14) Oral arm
15) Ovary
16) Perisarc
17) Retractor muscle
18) Rhopalium
19) Tentacle
20) Velum

PLATYHELMINTHES

Inside *Digital Zoology*

As you explore the Platyhelminthes on the *Digital Zoology* CD-ROM, don't miss these learning tools:

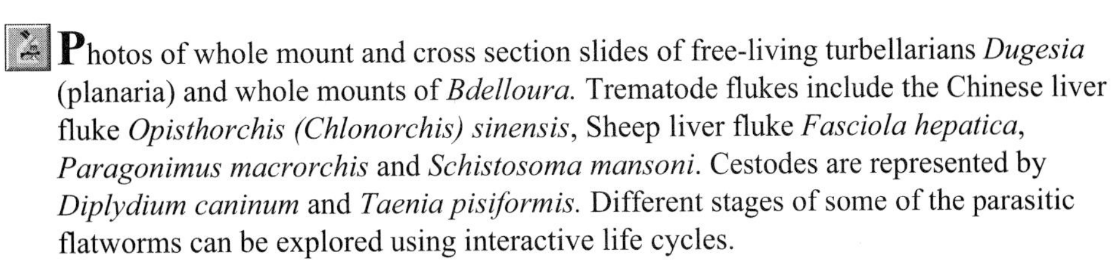

Photos of whole mount and cross section slides of free-living turbellarians *Dugesia* (planaria) and whole mounts of *Bdelloura.* Trematode flukes include the Chinese liver fluke *Opisthorchis (Chlonorchis) sinensis*, Sheep liver fluke *Fasciola hepatica*, *Paragonimus macrorchis* and *Schistosoma mansoni.* Cestodes are represented by *Diplydium caninum* and *Taenia pisiformis.* Different stages of some of the parasitic flatworms can be explored using interactive life cycles.

Video clips of flatworm movement showing gliding ciliary movement and use of musculature in *Dugesia* are available in addition to video clips showing the avoidance response and the pharynx.

Drag-and-drop quizzes on the key structures of the *Dugesia*, the Chinese liver fluke, and tapeworms.

Interactive cladogram showing the major events that gave rise to the four platyhelminth classes. A summary of the key characteristics of each class are combined with an interactive glossary of terms.

Defining Differences

Some of the differences described in the following sections are features that arise for the first time in the Platyhelminthes, and they define the phylum. Others, in the following list, are important for understanding how animals in the phylum function. Whichever is the case, you'll want to watch for examples of these in the various flatworm specimens that you will be examining.

- Bilaterally symmetry,
- triploblast acoelomates, at the
- organ-grade organization that have either
- free-living, or parasitic lifestyles.

Bilateral symmetry

Bilateral symmetry appears for the first time in the flatworms. The common assumption is that this type of symmetry is more advanced than the preceding **radial symmetry**, or **biradial symmetry.** (This is discussed in more detail in the previous chapter on Cnidaria.) Bilaterally symmetric animals have anterior and posterior ends, left and right sides, as well as dorsal and ventral surfaces. But being bilateral is more than just a new term to describe the parts of the animal. There are advantages to being able to sense where you're going, and there is increasing complexity of the nervous system and a concentration of sensory structures at the anterior end of the animal. It's a process referred to as **cephalization**, a first with the flatworms, but with a difference. In other cephalized animals, the mouth is located at the anterior end but that's not the case in flatworms, and their mouth is located on the ventral surface.

Describe the cephalization of the flatworm's nervous and sensory systems.

Triploblast acoelomates

The Platyhelminthes are the first **triploblastic** animals and have a third tissue layer, the **mesoderm**, added between the **ectoderm** and **endoderm** that cnidarians had. Flatworms, and all the other triploblasts, can now have muscles, formed from mesoderm. Platyhelminthes don't realize the full potential of the triploblast body plan and, unlike the other animals to follow, a flatworm's mesoderm is solid without a cavity, a **coelom**, inside. It's why they are referred as **acoelomate**. About the only way a solid animal can maintain an adequate surface-to-volume ratio is to be as flat as possible, and that's what these animals do. It explains how the phylum got its common name, flatworms.

Solid organisms have to depend on diffusion to move oxygen, nutrients, and metabolic wastes, and the same is true for flatworms. Like the cnidarians, the digestive system is incomplete, but branches throughout the body making sure that every cell is close to a supply of nutrients. The gut contents are mixed by the cilia that line it and also movements of the animal. The fluids in the gut and the surrounding environment are important for gas exchange and removal of metabolic wastes. Flatworms also have **protonephridia** for **osmoregulation** and are most obvious in freshwater species, that are **hyperosmotic** to the water around them. Protonephridia are also called flame cells, and they collect the interstitial fluid found between the cells and remove it from the animal.

 *How does a flame cell work?*

Organ-grade organization

Even with the limitations of being acoelomate, flatworms have developed complex organ systems including: nervous and excretory systems and a complicated reproductive system, easy to see in your lab specimens and in *Digital Zoology*. Flatworms are **monoecious**, and each animal has both male and female reproductive tracts — they're hermaphrodites, and that's why the reproductive system appears to be so complicated.

Compared to **asexual reproduction**, one of the advantages of **sexual reproduction** is that there is mixing of genetic material from two different parents, and this results in offspring that have a different genetic make up than their parents. Being monoecious doesn't mean that this advantage is lost. Flatworms, like all monoecious organisms, go to great lengths to be sure that they don't fertilize their own eggs.

The secret is that events of insemination: transferring the sperm; and fertilization, combining the sperm and egg, have been separated from each other. Here's how it works. In each flatworm, sperm is formed in the **testes**, then transported by the **vas deferens** (sperm ducts) to the s**eminal vesicles** where it is stored until the animal meets with another flatworm. When the two mate, each passes sperm from their own seminal vesicle to the other flatworm, who stores the sperm that they received in their **seminal receptacle**. Now that insemination is over, the two mating flatworms separate and head off on their own.

Some time later fertilization occurs. Eggs produced in the **ovaries**, pass down the **oviduct**, and are filled with nutritious yolk from the **vitelline glands**. The eggs then pass by the opening of the seminal receptacle and are fertilized using the stored sperm from the other flatworm. The result is that the flatworms that hatch from these fertilized eggs have genetic material from each of the parents and variation is possible.

 *What advantage could there be to being a monecious organism?*

Free-living and parasitic lifestyles

In the evolutionary sequence, flatworms are replaced by coelomate animals, and it's not unlikely that the new coelomates fed on flatworms. A flatworm gliding by on the substrate might be a tasty morsel for a potential coelomate predator, and living flatworms give us clues to how the early animals in the phylum survived this new threat to still be around today. They either defended themselves against predation or moved into the nice fluid-filled cavity of these new coelomate animals, living there instead.

Ancestral flatworms moved using cilia that beat in a layer of mucos that the epithelium lay down on the substrate. Any modifications of the mucous that were distasteful to would-be predators were advantageous, and flatworms still use this defensive strategy. In some flatworms **rhabdites** form in the epithelium, and when these rod-shape structures are discharged, they rapidly dissolve to make the distasteful mucous that protects the flatworm. One of the more unusual defenses is found in flatworms that feed in cnidarians. Cnidocytes that the flatworm consumes don't fire, and they're moved to the flatworm's epithelium and used for defense.

Another way that flatworms move is to stick to the substrate, and then use contractions of the longitudinal and circular muscles to move the body. Again, epithelial secretions are important in sticking to the substrate, and in parasitic flatworms adhesive glands are located in specific areas creating adhesive suckers used to stay attached to the host. In still other parasites, the epithelium secretes hooks used to hold the animal in place.

Adopting a parasitic existence is no easy task. To survive, a parasite can't kill its host, but has to be able to locate, and move into, a new host if the need arises. To do that, the number of parasites in the host needs to be increased. After all, if it's only a one in a thousand chance that the parasite can get to a new host, it may be out of luck if the one parasite that first found the host has only 500 offspring when it's time to move to a new host. Now, if there are 50,000 offspring, that's a different story. It's what the life cycles are about, increasing the numbers and getting to the next host. What makes the life cycles of parasitic flatworms stand out, especially those of the trematode flukes, is their complexity involving not one, but two and sometimes three, different animals.

 *What are the different stages in the fluke life cycle, and what does each contribute to the cycle?*

A Closer Look Inside *Digital Zoology* - Platyhelminthes

Turbellaria: Free-living flatworms Dugesia *and* Bdelloura

The flatworm *Dugesia*, which has the common name planaria, is the most often studied example of a flatworm in zoology courses. If living specimens aren't available in your lab, be sure and see the two different ways that *Dugesia* moves in the video clips in *Digital Zoology*.

The first way of moving is by using cilia, and it results in a smooth gliding motion across the substrate. In the second, which can best be described as a "stick and stretch" technique, adhesive glands on its ventral surface are used to stick to the substrate. With the anterior end stuck in place, longitudinal muscles contract and *Dugesia* shortens its body. It then sticks the posterior end in place. Once that's done, the anterior end is released from the substrate. The circular muscles contract causing the body to lengthen and push against the anchored posterior end moving the anterior end forward. Once again stick the anterior back down and repeat the sequence.

Whole mount slides give you a good look at the digestive tract in these animals, and you'll see how it branches throughout the body to supply nutrients to all the cells and tissues. *Dugesia* is a **triclad** flatworm because its gut has three branches. You'll find the pharynx on the ventral surface, although you may have to flip your microscope slide upside down to see it. A short video clip in *Digital Zoology* shows the pharynx. Other easy features to see on whole mounts are the **eye spots** and auricles, which are evidence of cephalization and form part of the nervous system.

Prepared slides of cross sections usually include sections through the animal anterior and posterior to the pharynx and through the pharyngeal region. Be sure to look at each of these, and try and match where the section comes from with the whole mount slide that you looked at. As you look at your cross section slides be careful that you don't confuse the lumen of the digestive tract or the pharynx with a body cavity. Remember that these animals are acoelomate.

 *How can you tell if your cross section is anterior to, posterior to, or through the pharyngeal region?*

Differences in the length of the cilia make it possible to tell the dorsal surface, with its shorter cilia, from the ventral surface, with its longer cilia. Look toward the edges of the section, and you should be able to see the adhesive glands that *Dugesia* uses to stick to the substrate. The circular and longitudinal muscles are easy to see along the edge of the body with a third type of muscle, the dorsoventral muscles, which extend from the dorsal to the ventral surface of the animal. These help to keep the flatworm flat as it contracts the other musculature.

If your lab doesn't have *Bdelloura* available, you might want to take a look at the whole mount of it in *Digital Zoology*. Preparations of this flatworm are more transparent than *Dugesia,* making it a lot easier to see the detail of the internal organs, especially the reproductive and nervous systems. *Bdelloura* is an ectoparasite found on the gills of horse shoe crabs. They stick to their host using a caudal sucker made of large adhesive glands similar to the ones found along the edges of *Dugesia*. When epidermal gland cells form a localized sticking structure like this, it's called a sucker, and we'll see more examples of this in the endoparasitic flatworms.

Trematoda: Trematode flukes

There are four different flukes available for you in *Digital Zoology*: the Chinese liver fluke *Opisthorchis (Chlonorchis)*, Sheep liver fluke *Fasciola*, *Paragonimus,* and *Schistosoma.* The full life cycle of the Chinese liver fluke is also available. Photos of whole mounts of all four adults will allow you to identify various internal structures.

The Chinese liver fluke, *Opisthorchis*, is used in many labs to demonstrate the trematode life cycle. Slides of the adults allow you to easily see the mouth and digestive tract, which forks into two branches just behind the mouth. Most of the animal is filled with reproductive tract. Paired testes occupy the posterior end of the animal and the female reproductive tract, the middle and anterior. The ovary is in the middle along with the seminal receptacle where sperm from another fluke is stored. The uterus winds its way to the genital pore, and you should be able to see shelled eggs inside. Yolk glands on the sides of the animal connect in the region of Mehlis' gland and the opening of the ovary.

In the Sheep liver fluke, *Fasciola*, the digestive tract makes it difficult to see many of the other structures. But if you look carefully you should be able to see most of the reproductive and digestive systems. Over all the organization is similar to that of the Chinese liver fluke with the male reproductive system occupying the posterior of the animal and the female in the anterior with the ovary near the middle. Vitelline glands are found down the sides. Try taking a close look at the anterior end of the animal so that you can differentiate what the digestive tract looks like from the reproductive system, which will help when you try and find the ovary. The uterus should be filled with shelled eggs, and posterior to it is the ovary. The vitelline duct carries yolk to the region of the ootype, an enlarged space in the reproductive tract, which also has openings from the ovary and seminal receptacle.

A third fluke that you'll find in *Digital Zoology* is *Paragonimus*. It's included in *Digital Zoology* because you'll have no trouble identifying the various parts of the internal anatomy with this species. Yolk glands, testes, ovaries, and the various tubes and ducts that connect them are easy to see in these almost transparent animals.

The last fluke we have for you is *Schistosoma*, which causes schistosomiasis. The adults of this species don't look like the other two. One of the big differences is that the sexes are separate in this species. Males and females also differ in their appearance with the males being the larger of the two and having a special groove that runs the length of his body, in which he holds a female.

In all four species be sure you can find both the anterior sucker and ventral sucker, characteristic of the trematodes.

Cestoda: Tapeworms

We have two different tapeworms for you to look at in *Digital Zoology*, and they closely resemble each other. In the dog tapeworm, *Dipylidium*, reproductive structures are paired with a gonopore on each side of the **proglottid**. In the Dog and cat tapeworm, *Taenia*, the structures aren't paired on each side. There are also morphological differences in the **scolex** that they use to attach to the intestinal wall of their vertebrate host.

The scolex is made up of hooks and suckers and behind is a a short neck that connects to the long chain of proglottids. New proglottids are added at the base of the neck, and the oldest ones are those farthest from the neck. Each proglottid is really just an egg-making machine, and when you examine the proglottids on the microscope slides, the reproductive tract is the most obvious feature. In younger proglottids you can see the testes, paired ovaries, and unpaired yolk gland. The vas deferens and oviduct both open in the gonopore. During mating sperm is passed out of the gonopore from the vas deferens, and into the gonopore, oviduct, and to the seminal receptacle of the other tapeworm. Be sure to look at a mature, **gravid,** proglottid to see how it is nothing more than a sac of eggs. Eggs don't pass down the oviduct and are released when the proglottid bursts.

Tapeworms live in the digestive tract of their vertebrate host and are able to absorb nutrients from the digestive fluids that surround them. Tapeworms don't have a digestive tract of their own.

How do tapeworms and flukes protect themselves from being digested or attacked by their host?

Cross-References to Other McGraw-Hill Zoology Titles

Integrated Principles of Zoology, 11th edition. C.P. Hickman, L.S. Roberts & A. Larson. Chapter 14.

Animal Diversity, 2nd edition. C.P. Hickman, L.S. Roberts & A. Larson. Chapter 7.

Zoology, 5th edition. S.A. Miller & J.P. Harley. Chapter 10.

Biology of the Invertebrates, 4th edition. J. Pechenik. Chapter 8.

Laboratory Studies in Integrated Principles of Zoology, 10th edition. C.P. Hickman, F. Hickman & L. Kats. Chapter 9.

General Zoology Laboratory Guide, 13th edition. C. Lytle. Chapter 9.

Structures Checklist

Here are some of the structures that you should be able to easily find in *Digital Zoology* and the specimens that you will be looking at in your lab. After reading your lab handout, you might want to add more and, depending on the equipment available in your lab, you might see more. As you study the material, you might also want to make some notes on how some of these structures looked or include a drawing in your lab notes. (Structures indicated by * may be hard to see.)

Dugesia

Preserved or whole mount

- ☐ Anterior branch of gut
- ☐ Auricles
- ☐ Eye spots
- ☐ Mouth
- ☐ Pharyngeal cavity
- ☐ Pharynx
- ☐ Posterior branches of gut
- ☐
- ☐
- ☐

Cross section

- ☐ Anterior branch of gut
- ☐ Cilia
- ☐ Circular muscle
- ☐ Diverticula of gut
- ☐ Dorsoventral muscle
- ☐ Epidermis
- ☐ Gastrodermis
- ☐ Longitudinal muscle
- ☐ Lumen of pharynx
- ☐ Pharyngeal cavity
- ☐ Pharynx
- ☐ Posterior branch of gut
- ☐ Rhabdites*
- ☐ Ventral nerve cords*
- ☐
- ☐

Additional structures

- ☐
- ☐
- ☐
- ☐

Bdelloura

Whole mount

- ☐ Adhesive disc
- ☐ Brain
- ☐ Diverticula of gut
- ☐ Eye spots
- ☐ Mouth
- ☐ Nerve cords
- ☐ Pharynx
- ☐ Seminal receptacles
- ☐ Sperm duct
- ☐ Testes
- ☐
- ☐
- ☐

Additional structures

- ☐
- ☐
- ☐
- ☐
- ☐
- ☐

Opisthorchis *(Chlonorchis)*

Adult whole mount

- ☐ Anterior testis
- ☐ Bladder
- ☐ Eggs
- ☐ Esophagus
- ☐ Excretory duct*
- ☐ Excretory pore
- ☐ Genital pore
- ☐ Intestine
- ☐ Mehlis' gland
- ☐ Mouth
- ☐ Ootype
- ☐ Oral sucker
- ☐ Ovary
- ☐ Pharynx
- ☐ Posterior testis
- ☐ Seminal receptacle
- ☐ Sperm ducts (vas efferens & vas deferens)*
- ☐ Uterus
- ☐ Ventral sucker (Acetabulum)
- ☐ Vitelline (Yolk) duct*
- ☐ Vitelline (Yolk) gland
- ☐
- ☐

Life cycle

- ☐ Eggs
- ☐ Miracidium
- ☐ Redia inside sporocyst
- ☐ Cercaria
- ☐ Metacercaria

Fasciola

Adult whole mount

- ☐ Anterior testis
- ☐ Eggs
- ☐ Esophagus
- ☐ Genital pore
- ☐ Intestine
- ☐ Mehlis' gland
- ☐ Mouth
- ☐ Ootype
- ☐ Oral sucker
- ☐ Ovary
- ☐ Penis*
- ☐ Pharynx
- ☐ Posterior testis
- ☐ Seminal receptacle
- ☐ Seminal vesicle*
- ☐ Sperm ducts (vas efferens & vas deferens)*
- ☐ Uterus
- ☐ Ventral sucker (Acetabulum)
- ☐ Vitelline (Yolk) duct*
- ☐ Vitelline (Yolk) gland
- ☐
- ☐
- ☐

Life cycle

- ☐ Eggs
- ☐ Miracidium
- ☐ Redia inside sporocyst
- ☐ Cercaria
- ☐ Cyst

Paragonimus

Adult whole mount

- ☐ Testes
- ☐ Eggs
- ☐ Esophagus
- ☐ Genital pore
- ☐ Intestine
- ☐ Mouth
- ☐ Ootype
- ☐ Oral sucker
- ☐ Ovary
- ☐ Pharynx
- ☐ Seminal receptacle
- ☐ Sperm ducts
- ☐ Uterus
- ☐ Ventral sucker (Acetabulum)
- ☐ Vitelline (Yolk) duct
- ☐ Vitelline (Yolk) gland
- ☐
- ☐
- ☐

Dipylidium *or* Taenia

Adult whole mount

- ☐ Cirrus*
- ☐ Excretory canal
- ☐ Genital pore
- ☐ Ootype
- ☐ Ovary
- ☐ Oviduct
- ☐ Seminal receptacle*
- ☐ Sperm duct (vas deferens)
- ☐ Testes
- ☐ Uterus
- ☐ Vagina
- ☐ Yolk gland
- ☐
- ☐
- ☐
- ☐
- ☐

Adult whole mount

- ☐ Cysticercal stage
- ☐ Gravid proglottid
- ☐ Immature proglottid
- ☐ Neck
- ☐ Rostrum with hooks
- ☐ Scolex
- ☐ Suckers on scolex
- ☐

Crossword Puzzle – Platyhelminthes

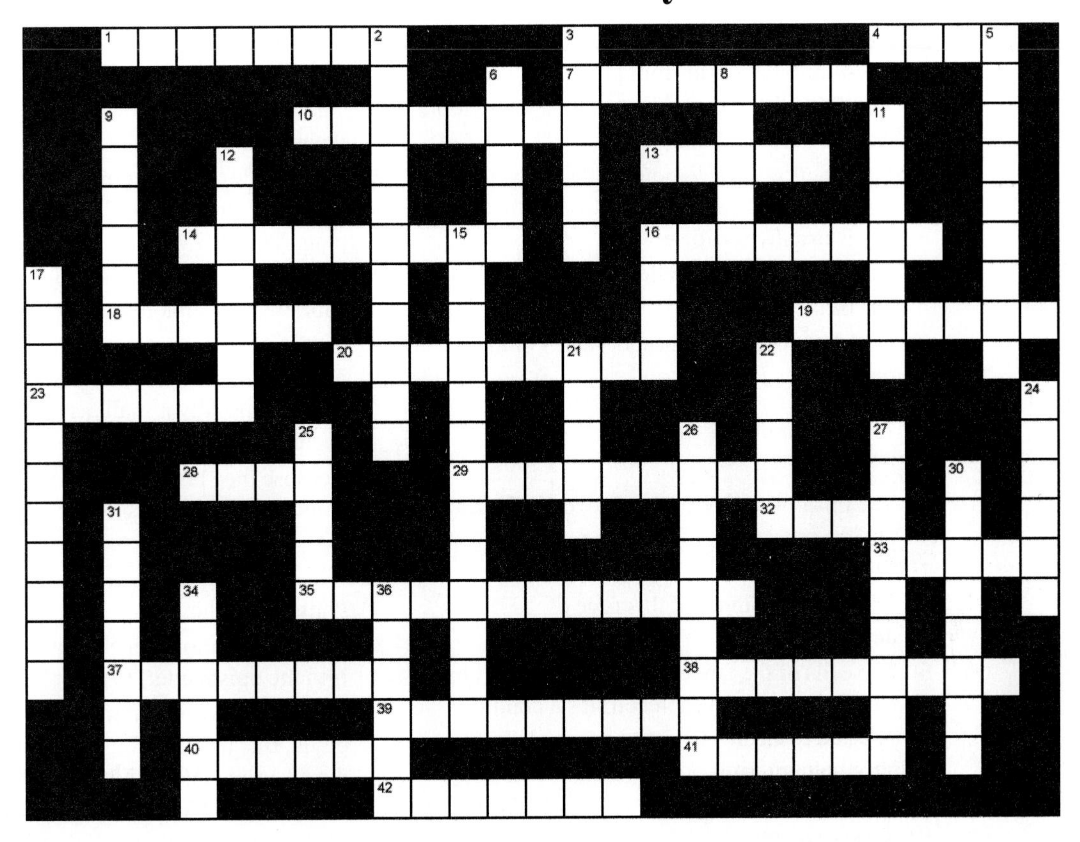

Puzzle solution. An interactive web based version of this puzzle, and its solution, are available on the *Digital Zoology* web site at www.mhhe.com/DigitalZoology/Students. With the interactive puzzle you can check to see if individual words or the whole puzzle is correct, and get hints for single letters.

Across

1 The special name for the protective body covering of parasitic flatworms. (8)

4 A flatworm's incomplete gut is missing this structure. (4)

7 These structures found on the sides of a free-living flatworm's head are chemosensory. (8)

10 This is the orientation of the outermost layer of muscles in free-living flatworms such as Planaria. (8)

13 Position of the epithelial cell bodies relative to the basement membrane in flatworm tegument. (5)

14 Another name for the yolk gland is this type of gland. (9)

16 The common name for the phylum Platyhelminthes. (8)

18 A tapeworm's head. (6)

19 A free living flatworm's mouth is located on the side of the animal. (7)

20 One of two types of hooks that you'll find on a monogenean fluke. (9)

23 The most mature segments of a tapeworm are loaded with eggs and referred to as being in this state or condition. (6)

28 The name of the anterior most sucker in a trematode. (4)

29 This organ system is missing in tapeworms. (9)

32 How many anterior branches are there in a polyclad flatworm's gut? (4)

33 Free-living flatworms use this sense to sample the chemical environment around them. (5)

35 Unlike cestodes and trematodes, the monogenean flukes are this type of parasite. (12)

37 These light sensitive structures make a planarian look cross-eyed. (3,5)

38 Embedded in the epidermis, these protect a free-living flatworm from predation. (9)

39 This flatworm class have acetabula; that's more than one acetabulum. (9)

40 The junction of the oviduct, and the opening of the seminal receptacle, a flatworm where the egg is usually fertilized. (6)

41 The nervous system of the flatworms looks like one of these. (6)

42 Females store their partners sperm in this receptacle. (7)

Down

2 The most primitive of the flatworm classes. (11)

3 The production of many cercaria from a single redia is an example of this type of amplification. (6)

5 The cells of the tapeworm tegument don't have cell walls between each cell. It's a condition referred to as this. (9)

6 The common name for a trematode. (5)

8 The body of the ancestral flatworm is covered with these. (5)

9 For cilia to work effectively the epidermis must first secrete this substance to cover the substrate that the flatworm glides across. (6)

11 Planaria feeds by using this structure to get food into its digestive tract. (7)

12 A flatworm with three main branches to its digestive tract is referred to as being this. (7)

15 The external opening of the protonephridia on the outer body wall of a flatworm. (13)

16 Number of classes in the phylum Platyhelminthes. (4)

17 These units repeat down the length of a tapeworm. (11)

21 The number of major muscle types, or layers, found in a free-living flatworm such as planaria. (5)

22 A flatworm in which the digestive system is a solid mass of endodermal cells, rather than a cavity. (5)

24 These flatworm parasites always have a mollusc, usually a snail, as part of their life cycle. (6)

25 Because they resemble a flickering candle, this name applies to the cells of the protonephridia. (5)

26 This type of symmetry is found in the phylum Platyhelminthes. (9)

27 The end of a monogenean fluke where you will find the opisthaptor. (9)

30 During the evolution of the different animal phyla, this tissue is found in the flatworms for the first time and is the reason they have muscles. (8)

31 This vital organelle lies below the plasma membrane of the tapeworm tegument and is protected. (7)

34 This cavity is missing in an acoelomate animal. (6)

36 The vas deferens transports gametes from this organ to the seminal vesicle. (6)

Self Test - Platyhelminthes

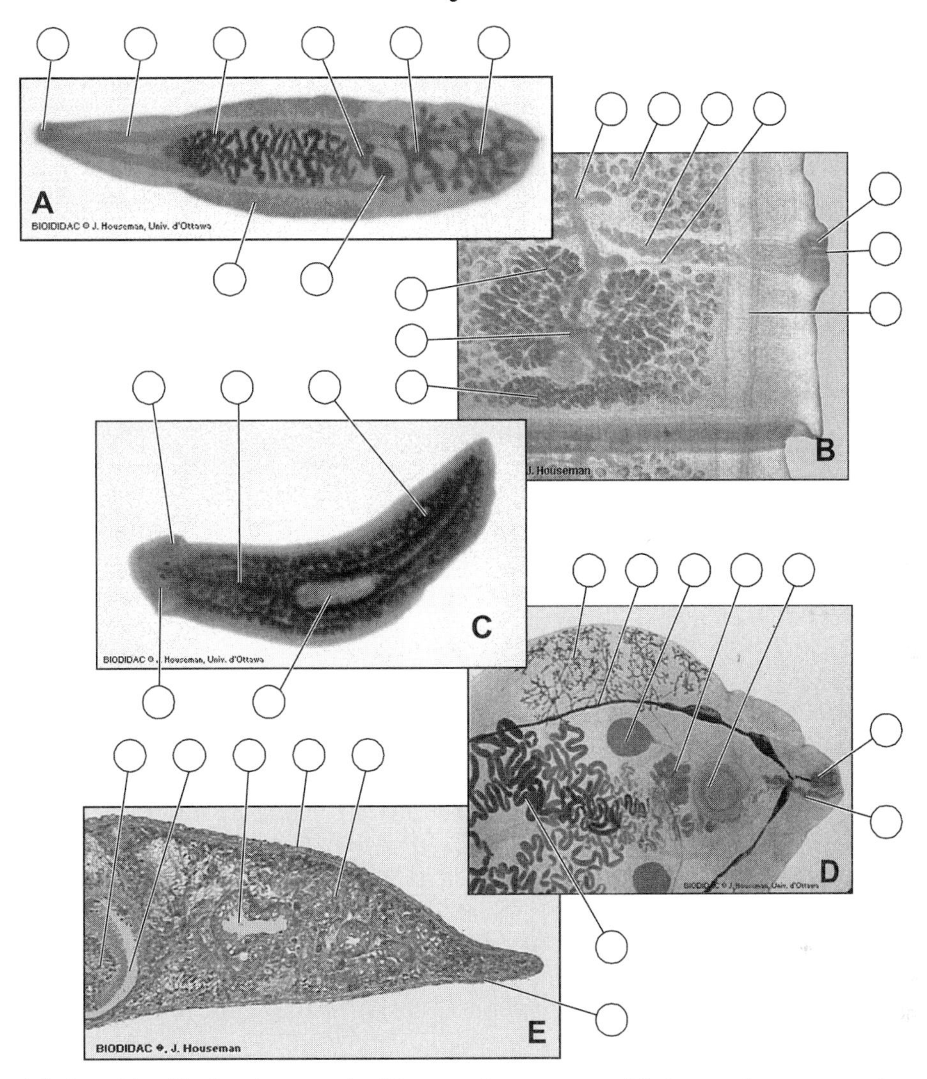

Use the following labels to identify the photographs. You may have to use a label more than once, and some labels may not be appropriate for the photographs. Answers are available on the Digital Zoology web site at www.mhhe.com/digitalzoology. Be sure to try the interactive Drag-and-Drop quizzes that are available on the *Digital Zoology* CD-ROM. A color version of this Self Test is available in the Adobe Acrobat version of the Student Workbook in the workbook folder on the *Digital Zoology* CD-ROM.

Specimens

A- *Opisthorchis* (*Chlonochis*)
B- *Diplydium*
C- *Obelia*
D- *Paragonimus*
E- *Dugesia*

Labels

1) Adhesive gland
2) Anterior branch of triclad intestine
3) Anterior testis
4) Auricle
5) Cirrus
6) Diverticula of the gut
7) Epidermis
8) Excretory canal
9) Eye spot
10) Genital duct
11) Gonopore
12) Intestine
13) Ootype
14) Oral sucker
15) Ovary
16) Oviduct
17) Parenchyma
18) Pharyngeal cavity
19) Pharynx
20) Posterior branch of triclad intestine
21) Posterior testis
22) Seminal receptacle
23) Testis
24) Uterus
25) Vas deferens
26) Yolk (Vitelline) gland

NEMATODA

Inside *Digital Zoology*

As you explore the nematodes on the *Digital Zoology* CD-ROM, don't miss these learning tools:

Photos of dissected specimens and cross sections of the Pig roundworm, *Ascaris lumbricoides*.

Drag-and-drop quizzes on the key structures of *Ascaris*.

Video clips showing the whiplike movement of *Turbatrix aceti*, the vinegar eel.

Interactive cladogram showing the major events that gave rise to the pseudocoelomate phyla, sometimes referred to as the phylum Aschelminthes. A summary of the key characteristics of each taxon are combined with an interactive glossary of terms.

Defining Differences

Some of the differences described in the following sections appear for the first time in the Nematoda, and they define the phylum. Others are important for understanding how animals in the phylum function, or they may be related to other pseudocoelomates. Whichever is the case, you'll want to watch for examples of these in the nematode specimens that you will be examining.

- A pseudocoelomate animal with a
- pseudocoel,
- longitudinal muscles, an outer
- cuticle,
- complete digestive tract, and special sensory
- amphids.

Pseudocoelomates

Should there be a phylum **Aschelminthes** that includes seven or eight **taxa** linked by **monophyletic** characters? Or are these taxa an artificial group, the **pseudocoelomates**, with nothing to link them together but a few **convergent traits**? The debate rages in zoological circles, and the taxonomic status of these animals, which only a few years ago were collectively referred to as the pseudocoelomates, is being revised once again. Things have essentially gone full circle from when the Aschelminthes were a phylum, to the term being dropped because the taxa in the phylum weren't related to each other. Now some biologists feel that these animals are linked by the presence of a **cuticle** and the absence of **cilia**, or **microvilli**, on their ectodermal surfaces and want to bring back the phylum Aschelminthes.

There's more involved in the debate than the validity of the Aschelminthes, and some scientists feel that molecular evidence requires even further revisions of the classification and that there be a whole new taxon, the **Ecdysozoa**. This would include the old pseudocoelomates, calling them once again the Aschelminthes, and the Arthropods. Monophyletic characters uniting the group are that all these animals molt and lack ectodermal cilia. While we wait for the dust to settle on this, *Digital Zoology* uses a conservative approach to the issue and places the pseudocoelomates as a group outside the main cladogram. It's a decision that can easily be criticized but, so could any of the alternatives as each has its supporters and critics.

✍ *In this new taxonomic scheme make a list of the phyla that would be included in the Ecdysozoa.*

That having been said, who are the pseudocoelomates? They're a group of animals that share a number of characteristics that include: being microscopic in size; having a pseudocoel, an outer cuticle (more on these two later), and **protonephridia** for excretion and osmoregulation; and exhibiting **eutely**.

We've seen protonephridia before, in flatworms, where they were also called flame cells. Powered by the cilia found inside the protonephridia, these closed-ended tubes create an ultrafiltrate of the surrounding fluid.

The body of a eutelic animal always has the same number of cells, and the number is species specific. Other examples of **eutely** can include animals where certain organs, or tissues, are always made of the same number of cells. If their tissues are **syncytial,** there are always the same number of nuclei in the tissue and again that number is species specific. If there are this many cells, it's species A; if it's that many, it's species B. A little simplistic but it gives you the idea.

You'll find all but one of these pseudocoelomate traits in nematodes. Instead of protonephridia, they have unusual glandular cells, **renette cells,** which have a presumed role in excretion and osmoregulation.

Pseudocoel

The pseudocoelomates get their name from their body cavity, the pseudocoel, which isn't the same as a true body cavity, the **eucoelomate** condition. Unlike a true coelom, a pseudocoel isn't completely lined with mesoderm, and a **peritoneum** doesn't completely surround the cavity. Instead, the **mesoderm** of a pseudocoel is found adjacent to only structures formed from **ectoderm** and not those derived from **endoderm**. A consequence of this, is pseudocoelomate animals don't have any musculature lining the gut, nor do they have mesenteries holding the internal organs in place.

✍ *How do pseudocoelomates swallow their food?*

One of the problems of using the pseudocoel as a unifying character for all these animals is that not all of these animals have body cavities. In some specimens, what was once thought to have been a pseudocoelomic space was an artifact from how the samples had been prepared or the specimens preserved, and these animals are actually acoelomate. At one point biologists thought they had it figured out. They thought the pseudocoel was the remains of the **blastocoel** that persisted in the adult. In some animals it remained as an open space, and in others it filled in again. But, even that is open to debate. One thing there is no debate about, is that nematodes have a body cavity, and that it fits the traditional definition of a pseudocoel.

Longitudinal muscles

Nematodes have only **longitudinal muscles**, and with a single fluid-filled pseudocoel acting as a **hydrostatic skeleton**, any contraction of the musculature alters the shape of the whole animal. It's the reason why movement in nematodes is characteristically whiplike, with wavelike undulations running the length of the body. One of the unusual features of nematode muscles is that it's the muscles that send cytoplasmic arms to connect with the ventral or dorsal nerve cords. That's the complete opposite of what you see in other animals where the nerves extend cytoplasmic extensions, axons, to the muscles.

Cuticle

Nematodes, like a variety of other animals, have a nonliving outer covering, the **cuticle**, produced by the **epidermis** underneath. In most invertebrates that don't have a cuticle, the outer epidermal surface is covered in cilia, which have a variety of functions. The presence of an outer cuticle changes that, and not surprisingly, wrapping a nonliving outer covering over the epidermal surface means that cilia can't function. They have disappeared from the surface of animals with cuticles.

Invertebrates have cuticles made of either **collagen**, protein polymers, or **chitin**, carbohydrate polymers. A nematode's cuticle is collagenous and made up of three layers that form a mesh, or lattice, of fibers. The collagen fibers are inflexible but, the way that they're organized in the mesh allows the body of the nematode to bend and flex. Think of the protective gate placed at the top of a flight of stairs to keep small children from accidentally falling down the stairs. The wooden bars in the gates are inflexible, like the collagen fibers, but the gate can change its shape becoming long and narrow or short and high. That's the secret to the hydrostatic skeleton and cuticular covering of nematodes.

Complete digestive tract

Although where and how nematodes and other pseudocoelomates fit into the phylogeny of the animals may be uncertain, in most zoology courses they are often the first animals encountered that have a **complete digestive tract** that includes a **mouth** at the anterior end and an **anus** at the other end of the animal. It may not seem like a big thing but it was, and here's why.

With an **incomplete digestive tract**, any undigested food must be removed from the system by the opening used for getting new food into it, the mouth. Imagine an animal with a partially digested meal in an incomplete gut when it encounters more food. There are two ways to deal with this, either remove the old food from the gut and then consume the new meal, or consume a smaller meal and add it to what's already in the digestive system.

It's a problem that animals with a complete gut don't have. They can feed whenever food is available. More food can be added at the anterior end, as partially digested food is moved back toward the posterior end of the animal and ultimately out the anus. Food can now be processed in a linear sequence, and whenever an opportunity to feed occurs, it can happen. This type of organization is also described as a tube-within-a-tube digestive system, and it allows for regions of specialization to occur in the digestive tract. These can include ways to grind and mix the food and specialized regions where different digestive conditions, such as acidity or alkalinity, can be varied.

 *There is another advantage to having a tube-within-a-tube construction that other animals with a complete digestive tract have, that nematodes don't have. What is it?*

Amphids

Nematodes have a pair of unique sensory structures at the anterior end of the body called **amphids**. They open to the outside through a pore, and an amphid pouch is found at the base of a short duct. These are sensory structures and a nerve connects them to the brain that surrounds the pharynx. Inside, the amphids are ciliated, and it's the only place that you'll find cilia in the nematode. Until cilia were found here, it was assumed that nematodes had completely lost their cilia.

A Closer Look Inside *Digital Zoology* - Nematodes

Ascaris*: Pig roundworm*

If you've been reading about nematodes, one of the things you've been told is that they're supposed to be microscopic in size, and that's certainly not the case with the *Ascaris*, the Pig roundworm. In part, this is because *Ascaris* is a parasitic nematode that lives in the digestive tract of its host, and presumably their size is in some way related to their parasitic life cycle. Whatever the reason, it works well for zoology students. It's much easier to see the nematode characteristics in an animal this size than one that is microscopic.

In the preserved specimens, the most obvious difference between the male and female roundworm is the hooked posterior end of the male, where the genital pore is located. You don't see that in the female, and her genital pore is midway down the body. When the two worms mate the male's body is at right angles to the female, and he uses the spicules at the tip of his body to hold on to the female and guide sperm into the female genital pore.

Is there any other way that you could tell the difference between the sexes?

Before you dissect your specimen take a close look at it under the dissecting microscope, and be sure to see the anterior tip of the animal and the surface of the body wall. If you can't see the structures on the anterior end of your roundworm, take a look inside *Digital Zoology*. The mouth is surrounded by three lips and a pair of pores that lead to the amphids. This combination of lips and the position of the amphid pores, is how you can tell the difference between the dorsal and ventral surfaces of the animal, with the single lip above the amphids being on the dorsal side and the other two lips below the amphids being on the ventral side of the animal.

Although at first glance you might think that the surface of the worm is smooth, with a closer look under the microscope you'll see that it's grooved and covered in ridges. Although not as important in *Ascaris* these ridges are part of how nematodes move.

The insides of a dissected specimen are deceptively simple, and you won't appreciate the complexity until you combine what you're seeing in the dissecting tray with the cross section slides. Most of the pseudocoel is filled with the reproductive tract, which looks like a tangled mass of threads. It is a tubular gonad with one end unattached and floating free in the pseudocoel, while the other end of the gonad is anchored to the genital pore. It's a long tube that winds back and forth inside the body cavity, and as it does so, its diameter gradually increases. Along its length are parts where the germ cells produce the gametes, other parts where gametes develop and mature, and finally, a place to store them before being released. Before you think you've got it sorted out, there's one last twist to the set up. The females have two of these tubular gonads and males only one.

You should be able to find where the tubes end, but the small start of the gonad may be hard to find. Other than seeing the changes in diameters, you can't tell the difference between the regions of gamete formation, maturation and storage. That's where the cross section slides come in. In the female, the solid, wagon-wheel-like section, is the ovary. The muscular oviduct without shelled eggs follows, and the paired uteri, with shelled eggs is the largest structure you can see in the cross section. In males it's harder to tell the differences and the size of the tube is your best guide. In most animals you would be able to identify the seminal vesicle because you would see the tails on the sperm stored there. But one of the unusual things about nematodes is that they have amoebic sperm!

Other things to match up between your dissection and the cross section are the muscles, the position of the lateral line, and the nerve cords (easier to see in the cross section). The gut is a straight tube that runs the length of the animal, and when you look at it in cross section you'll see there are no surrounding muscles.

Turbatrix: *Vinegar eel*

If available, prepare a wet mount of the vinegar eel and watch how these animals move. If live specimens aren't available, video clips can be found in *Digital Zoology*. As discussed previously, the longitudinal muscles work in combination with the pseudocoel, which acts as a hydrostatic skeleton. Any change in shape of the body wall changes the shape of the whole animal. The result is the whiplike movement that you see. It doesn't look like an efficient way to get around but remember, this movement occurs in substrate and the winding movement of the body is used to move between granules of the surrounding substrate. Remember those ridges that we mentioned in the cuticle? They're important in helping the nematode push against the substrate as the nematode moves.

Cross-References to Other McGraw-Hill Zoology Titles

Integrated Principles of Zoology, 11th edition. C.P. Hickman, L.S. Roberts & A. Larson. Chapter 15.

Animal Diversity, 2nd edition. C.P. Hickman, L.S. Roberts & A. Larson. Chapter 8.

Zoology, 5th edition. S.A. Miller & J.P. Harley. Chapter 11.

Biology of the Invertebrates, 4th edition. J. Pechenik. Chapter 16.

Laboratory Studies in Integrated Principles of Zoology, 10th edition. C.P. Hickman, F. Hickman & L. Kats. Chapter 10.

General Zoology Laboratory Guide, 13th edition. C. Lytle. Chapter 10.

Structures Checklist

Here are some of the structures that you should be able to easily find in *Digital Zoology* and the specimens that you will be looking at in your lab. After reading your lab handout, you might want to add more and, depending on the equipment available in your lab, you might see more. As you study the material, you might also want to make some notes on how some of these structures looked or include a drawing in your lab notes. (Structures indicated by * may be hard to see.)

Ascaris

Preserved specimen - intact and dissected

- ☐ Amphid pores
- ☐ Copulatory spicules
- ☐ Intestine
- ☐ Lateral line
- ☐ Lips
- ☐ Mouth
- ☐ Ovary
- ☐ Oviduct
- ☐ Pharynx
- ☐ Seminal vesicle
- ☐ Sperm duct
- ☐ Testis
- ☐ Uterus
- ☐ Vagina
- ☐
- ☐
- ☐

Cross section

- ☐ Cuticle
- ☐ Dorsal nerve cord
- ☐ Epidermis
- ☐ Intestine
- ☐ Lateral line (Renette cells)
- ☐ Longitudinal muscle
- ☐ Ovary
- ☐ Oviduct
- ☐ Seminal vesicle
- ☐ Sperm duct
- ☐ Testis
- ☐ Uterus
- ☐ Ventral nerve cord
- ☐
- ☐
- ☐

Additional structures

- ☐
- ☐

Crossword Puzzle – Nematoda

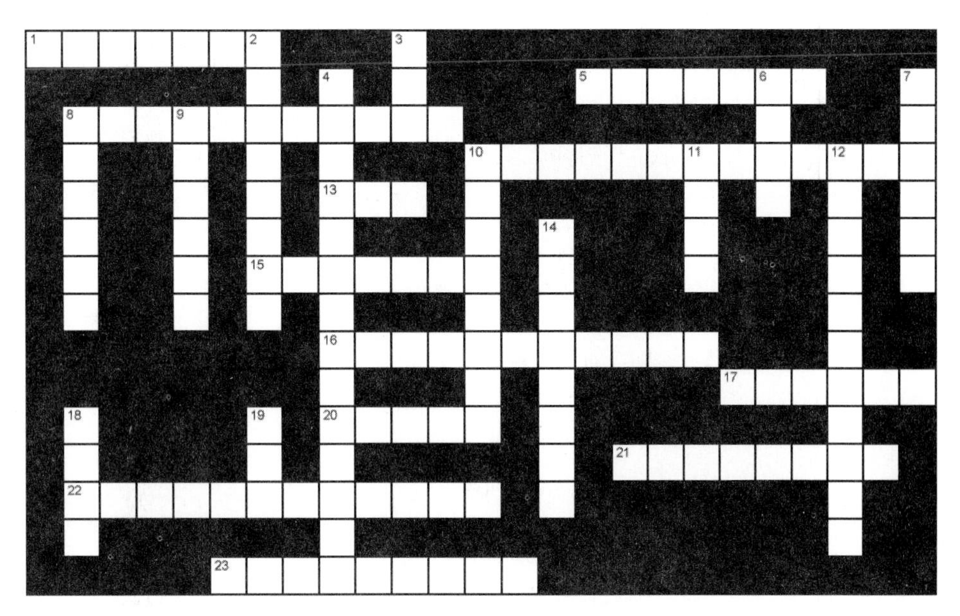

Puzzle solution. An interactive web based version of this puzzle, and its solution, are available on the *Digital Zoology* web site at www.mhhe.com/DigitalZoology/Students. With the interactive puzzle you can check to see if individual words or the whole puzzle is correct, and get hints for single letters.

Across

1 All pseudocoelomates have an outer nonliving body covering referred to as this. (7)

5 This muscular structure helps a nematode swallow its food. (7)

8 The internal organs of a nematode float free in the body cavity because there are none of these to anchor them in place. (11)

10 Pseudocoelomates were once referred to as this phylum. (13)

13 In a nematode there are this many ovaries. (3)

15 These cells in a nematode are thought to be osmoregulatory. (7)

16 This fluid filled cavity forms the hydrostatic skeleton of a nematode. (11)

17 Unlike most male animals which have two, male nematodes have only one of these. (6)

20 Animals in the phylum Nematoda are commonly called this type of worm. (5)

21 Nematodes have this type of a digestive system. (8)

22 Nematodes have only these muscles. (12)

23 Ascaris is an unusually large nematode and this results from its adaptation to this type of life. (9)

Down

2 The pharynx is actually formed from this germ layer and that's why it has muscle lining it. (8)

3 The number of muscle layers in a nematode. (3)

4 Nematodes don't have this osmoregulatory structure found in other pseudocoelomates. (14)

6 The total number of pseudocoelomate phyla. (4)

7 Extensions of these cells connect the nervous and muscular systems of a nematode. (6)

8 In nematodes the opening to the female genital system is found in this region of the body. (6)

9 Many species of pseudocoelomates demonstrate this where all the animals in the same species have the same number of cells (6)

10 The type of movement seen in nermatode sperm. (8)

11 The curl on the posterior end of a nematode identifies its gender as this. (4)

12 The fluid in the pseudocoel is used as this type of skeleton for locomotion. (11)

14 In a pseudocoelomate this tissue layer doesn't completely line the body cavity. (8)

18 Nematodes and other pseudocoelomates are a good example of a ____phyletic group. (4)

19 In a nematode the endoderm forms this structure. (3)

Self Test – Nematoda

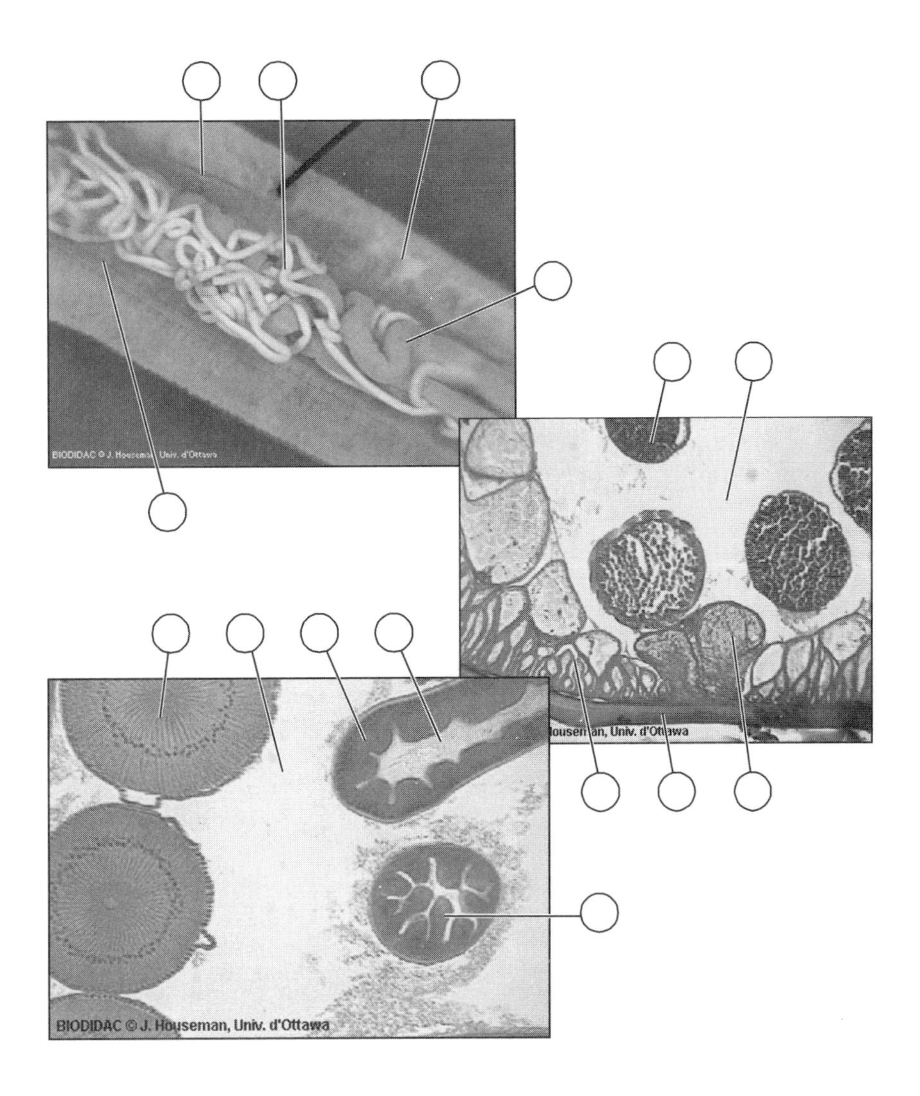

Use the following labels to identify the photographs. You may have to use a label more than once, and some labels may not be appropriate for the photographs. Answers are available on the *Digital Zoology* web site at www.mhhe.com/digitalzoology. Be sure to try the interactive Drag-and-Drop quizzes that are available on the *Digital Zoology* CD-ROM. A color version of this Self Test is available in the Adobe Acrobat version of the Student Workbook in the workbook folder on the *Digital Zoology* CD-ROM.

Specimen

Ascaris

Labels

1) Cuticle
2) Gut lumen
3) Gut wall
4) Lateral line
5) Longitudinal muscle
6) Ovary
7) Oviduct
8) Pseudocoel
9) Testis
10) Uterus

ANNELIDA

Inside *Digital Zoology*

As you explore the annelids on the *Digital Zoology* CD-ROM, don't miss these learning tools:

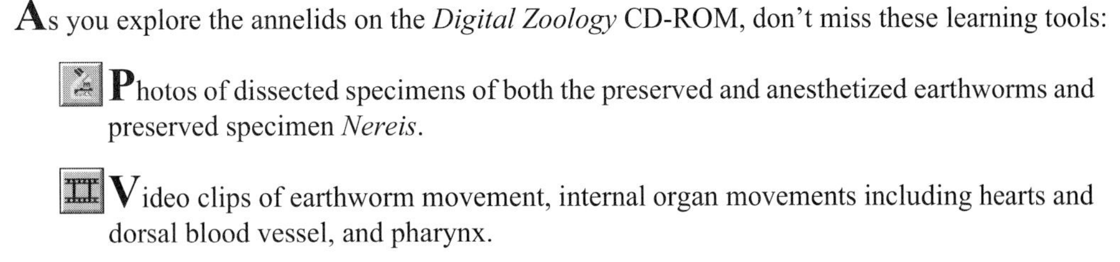

Photos of dissected specimens of both the preserved and anesthetized earthworms and preserved specimen *Nereis*.

Video clips of earthworm movement, internal organ movements including hearts and dorsal blood vessel, and pharynx.

Drag-and-drop quizzes on the key structures of the earthworm and *Nereis.*

Interactive cladogram showing the major events that gave rise to the three annelid classes.

Defining Differences

Some of the differences described in the following sections are features that arise for the first time in the Annelida, and they define the phylum. Others are important for understanding how animals in the phylum function. Whichever is the case, you'll want to watch for examples of these in the various annelid specimens that you will be examining.

- Metamerisation, a
- a closed circulatory system; and
- some new forms of locomotion.

Metamerisation

Unlike molluscs, where the coelom has been reduced and replaced with a hemocoel, the annelids retain a large and spacious body cavity. But annelids have done something a little different with the coelom. To become larger animals by repeating the coelomic unit, end to end, adding the new segments between the mouth and the anus. The proper term for the process is **metamerisation**, each of the segments is a **metamere**, and this is what give the annelids their unique segmented look.

In the ancestral annelid each metamere was a carbon copy of the one adjacent to it. Any structures formed from the **mesoderm**, and overlying **ectoderm**, were duplicated in each segment making each **serially homologous** to those adjacent to it. You could almost think of each segment as a complete independent organism with its own excretory system, circular and longitudinal muscles to work the fluid filled coelomic fluid as a **hydrostatic skeleton**, and a ganglion to coordinate events in the metamere. There were also paired **metanephridia** to filter coelomic fluid and remove metabolic wastes, as well as paired gonads on the septal wall of each metamere. As you work with your specimens and look through *Digital Zoology* be sure to make a note of metamerically arranged structures that you see both externally and internally.

✍ *Can you think of some of the advantages of having a body cavity?*

Closed circulatory system

With the metameric plan the advantage of the linear processing of ingested food made possible by a **complete gut** was now a liability. In other **eucoelomates** the coelomic fluid could be used to transport nutrients and wastes but in annelids the metameres are separated by **septa**. Coelomic fluid in adjacent segments doesn’t mix between segments. It couldn't be used to transfer nutrients between metameres located next to parts of the gut that had become specialized for nutrient absorption and those next to regions where absorption did not occur. The development of the **closed circulatory system** in combination with metamerism was important in ensuring that nutrients were evenly distributed throughout the ancestral annelid.

With the coelomic fluid no longer having a major role in transporting nutrients, oxygen, and metabolic wastes, one of it's other important roles, as a hydraulic medium, took on new importance for the annelids.

✍ *What are some of the different functions of coelomic fluid?*

Burrowing - a new form of locomotion

In the ancestral annelid each metamere was hydrostatically isolated from the one next to it by the septal walls between the segments. Individual metameres could be lengthened, with the associated decrease in diameter, or shortened, with an increase in length, without affecting the shape of the adjacent segment.

All of this allowed annelids to be the first animals to burrow into the organic sediments of the ancient oceans — a new form of locomotion. But there was more to this than moving through the sediments. They didn't push the sediments aside as they burrowed — they swallowed them, becoming the first substrate feeders.

✍ *Substrate feeding is one of a number of different feeding strategies. What are some of the others?*

This ancestral characteristic of segmentation has been modified and only in the most primitive annelids is near perfect **serial homology** still visible down the length of the animal. In more advanced forms, regions of specialization have occurred and **septa,** which separated each segment, often disappear. In the leeches the body cavity is no longer as spacious and has once again become filled with tissue.

Increased activity associated with a new form of locomotion also made demands on other systems. It required changes in the nervous and sensory systems to coordinate movement. The circulatory system is more elaborate, as would be expected to supply the increased metabolic demands associated with this enhanced mobility and the necessity of transporting nutrients from the digestive tract to each of the isolated metameres. Increased activity increases metabolism, which in turn requires a more efficient means to deal with metabolic wastes. As a consequence the excretory system has become elaborate in annelids. Changes in the excretory system have also permitted the invasion of the terrestrial habitat but also set the limits for this phylum's success within this environment. Terrestrial worms have to absorb oxygen across their body walls.

A Closer Look Inside *Digital Zoology* – Annelids

Polychaeta: Nereid worm

The marine worm *Nereis* is often used to introduce the annelids because many of the ancestral annelid characters can be seen in the animal. But be careful as you make your observations. Remember that the ancestor to the annelids burrowed and there were no **parapodia**, something unique to the polychaetes. Bundles of **setal hairs** helped the ancestral annelid anchor its body in the burrow. Only later did lateral extensions of the body wall form parapodia, allowing the worm to move across the surface — the beginnings of the polychaetes and yet another way to get around. Take a close look at the parapodia and you'll see how they have their own musculature, oblique muscles, and strengthening **acicula**.

As the annelid moves across the substrate it needs to know where its going, and the head has a variety of different sensory structures important for its predatory existence: eyes, jaws and sensory tentacles are the most obvious. Internally each coelomic space is still isolated from the next and other than **metanephridia**, ganglia, nerve cord, and blood vessels, there really isn't much to see inside. Cross sections show how the muscles are arranged in four bands down the length of the segment, and this relates to how the animal moves. Be sure to look at the simple arrangement of the blood vessels, ventral nerve cord, and the simple tubular gut. Try and match the structures that you see in the photos of the cross sections with the pictures of the dissected specimens.

Although burrowing may not have been ancestral to the polychaetes, there are some worms in the class that do just that. Polychaetes are often grouped into two types, free swimming (errant), and sedentary. *Nereis* is an errant polychaete. For sedentary polychaetes the parapodia may get in the way and they are often lost, or modified, for other functions. Two good examples of these changes are pumping water through the burrow and acting as gills. There also may be regional specialization down the length of the animal with parapodia doing different things depending on where they are found on the body. A great example is the Parchment worm, *Chaetopterus*.

Oligochaeta: The earthworm

The earthworm gives us a chance to see the burrowing body plan that was ancestral to the phylum. Unlike polychaetes, which used muscles arranged in bands for movement, in earthworms they are arranged in layers, with an outer circular and inner longitudinal muscles. Depending on which is contracting, the whole segment is either lengthened or shortened. The numbers of setae are reduced and two pairs are found on each side of the animal. Cross section slides will show you these and their musculature.

As you look at the external and internal features of the earthworm, remember that this is a highly specialized animal that has conquered the terrestrial environment. Just as we saw in the polychaetes, there are now regions of specialization where the ancestral pattern of identical segments has disappeared. Externally the most obvious example of this is the clitellum. Internally the gonads are permanent and only found in a few of the anterior segments. The reproductive system is further complicated because these animals are **monoecious** (hermaphrodite) organisms. Each animal has structures to store its own sperm (**seminal vesicle**) prior to mating and needs a mechanism to pass it to another worm during mating, while simultaneously receiving the other's sperm and storing it (**seminal receptacle**) for later. The secret to monoecism is that the events of sperm transfer and fertilization are separate.

 *Why is it important that the events of sperm transfer and fertilization be separated in a monoecious organism?*

The circulatory system of the earthworm is more complex than that of the polychaetes, and five blood vessels that connect dorsal and ventral blood vessels have been enlarged and referred to as hearts. The surface area of the gut has been enlarged by an inner fold, the **typhlosole**, and in the anterior there are parts of the gut that have become specialized. The gut and many of the blood vessels, are covered in **clorogogen tissue** that is involved in intermediary metabolism.

Cross-References to Other McGraw-Hill Zoology Titles

Integrated Principles of Zoology, 11th edition. C.P. Hickman, L.S. Roberts & A. Larson. Chapter 17.

Animal Diversity, 2nd edition. C.P. Hickman, L.S. Roberts & A. Larson. Chapter 10.

Zoology, 5th edition. S.A. Miller & J.P. Harley. Chapter 13.

Biology of the Invertebrates, 4th edition. J. Pechenik. Chapter 13.

Laboratory Studies in Integrated Principles of Zoology, 10th edition. C.P. Hickman, F. Hickman & L. Kats. Chapter 12.

General Zoology Laboratory Guide, 13th edition. C. Lytle. Chapter 12.

Structures Checklist

Here are some of the structures that you should be able to easily find in *Digital Zoology* and the specimens that you will be looking at in your lab. After reading your lab handout, you might want to add more and, depending on the equipment available in your lab, you might see more. As you study the material, you might also want to make some notes on how some of these structures looked or include a drawing in your lab notes. (Structures indicated by * may be hard to see.)

Earthworm

External anatomy

- ☐ Anus
- ☐ Clitellum
- ☐ Female pores*
- ☐ Male pores
- ☐ Mouth
- ☐ Peristomium
- ☐ Prostomium
- ☐ Pygidium
- ☐ Seminal groove
- ☐ Setae
- ☐
- ☐
- ☐
- ☐

Internal anatomy

- ☐ Aortic arches (hearts)
- ☐ Calciferous gland
- ☐ Chloragogen tissue
- ☐ Crop
- ☐ Dorsal vessel
- ☐ Esophagus
- ☐ Gizzard
- ☐ Intestine
- ☐ Metanephridium
- ☐ Nerve cord
- ☐ Ovaries*
- ☐ Pharynx
- ☐ Pharynx musculature
- ☐ Seminal receptacles
- ☐ Seminal vesicles
- ☐ Septal walls
- ☐ Subneural vessel
- ☐ Testes*
- ☐ Ventral vessel
- ☐
- ☐
- ☐

Cross section

- ☐ Chlorogogen tissue
- ☐ Circular muscle
- ☐ Coelomic space
- ☐ Cuticle
- ☐ Dorsal vessel
- ☐ Giant axons
- ☐ Gut lumen
- ☐ Gut wall
- ☐ Longitudinal muscle
- ☐ Metanephridia
- ☐ Setae
- ☐ Setal musculature
- ☐ Subneural vessel
- ☐ Typhlosole
- ☐ Ventral nerve cord
- ☐ Ventral vessel
- ☐
- ☐
- ☐

Additional structures

- ☐
- ☐
- ☐
- ☐
- ☐

Nereid worm

External anatomy

- ☐ Anus
- ☐ Dorsal cirrus
- ☐ Eye
- ☐ Jaws*
- ☐ Mouth
- ☐ Neuropodium
- ☐ Notopodium
- ☐ Parapodia
- ☐ Peristomial cirri
- ☐ Peristomium
- ☐ Prostomial palp
- ☐ Prostomial tentacles
- ☐ Prostomium
- ☐ Pygidium
- ☐ Setae
- ☐ Ventral cirrus
- ☐
- ☐
- ☐

Internal anatomy

- ☐ Digestive caecum
- ☐ Dorsal vessel
- ☐ Intestine
- ☐ Jaws
- ☐ Longitudinal muscles
- ☐ Metanephridia
- ☐ Nerve cord
- ☐ Pharynx
- ☐ Septal walls
- ☐ Ventral vessel
- ☐
- ☐
- ☐
- ☐

Cross section

- ☐ Acciculum
- ☐ Coelomic space
- ☐ Dorsal vessel
- ☐ Gut lumen
- ☐ Gut wall
- ☐ Longitudinal muscles
- ☐ Neuropodium
- ☐ Notopodium
- ☐ Oblique muscles
- ☐ Setae
- ☐ Ventral nerve cord
- ☐ Ventral vessel
- ☐
- ☐

Additional structures

- ☐
- ☐
- ☐
- ☐
- ☐

Crossword puzzle – Annelida

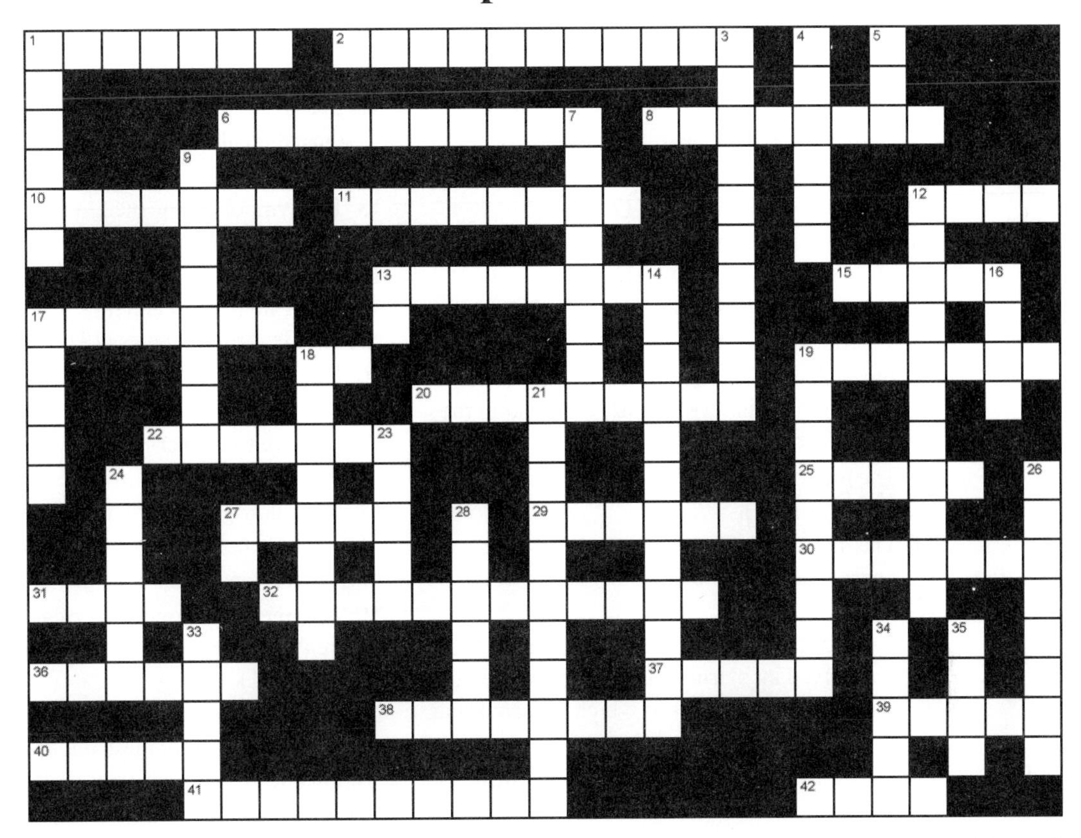

Puzzle solution. An interactive web based version of this puzzle, and its solution, are available on the *Digital Zoology* web site at www.mhhe.com/DigitalZoology/Students. With the interactive puzzle you can check to see if individual words or the whole puzzle is correct, and get hints for single letters.

Across

1 In leeches, the coelomic cavity has been reduced to spaces referred to as these. (7)

2 Parapodia are involved in locomotion and this function. (11)

6 Although this is the most anterior part of an annelid, it's not a true segment. (10)

8 In polychaete worms that demonstrate epitoky, the atoke segments are at this end of the animal. (8)

10 These highly modified parapodia create and hold onto the mucous bag in the filter feeding polychaete Chaetopterus. (7)

11 Metameres in a worm are also called these. (8)

12 Any part of the digestive tract involved in storing food prior to digestion; it's particularly large in leeches. (4)

13 The type of fluid you would find in both the metanephridia and coelomoducts. (8)

15 These structures on the funnel of the metanephridia pump the coelomic fluid into the nephridia of an annelid. (5)

17 These chitinous rods help support the parapodia of marine worms. (7)

18 Would you find hearts in the circulatory system of a marine worm? (2)

19 In monoecious annelids, sperm received during mating is stored in this receptacle. (7)

20 Like an earthworm, leeches have this structure, which secretes a protective case that surrounds their fertilized eggs. (9)

22 In polychaete worms that demonstrate epitoky, you would find these in the epitoke segments. (7)

25 The _____ esophageal ganglion is also called the brain (prefix). (5)

27 Blood in the dorsal blood vessel of a worm moves toward this part of the body. (5)

29 Both earthworms and leeches place their fertilized eggs in one of these. (6)

30 Describes a leech's the body cavity size compared to other worm. (7)

31 An earthworm has this number of paired setae on each body segment. (4)

32 Not all the segments may look the same in an annelid. What is this condition? (12)

36 Ingested calcareous rocks and stones are dissolved by this type of pH condition in the intestine of an earthworm. (6)

37 The number of pairs of seminal vesicles in an earthworm. (5)

38 The main branches that support the feathery fans of the filter-feeding polychaete, fan or plume worms. (8)

39 Because the chlorogogue (chlorogogen) tissue is involved in intermediary metabolism, it is, in some ways, analogous to which human organ? (5)

40 The supraesophageal ganglion of a worm is also called a this. (5)

41 This structure increases the surface area of an earthworm's digestive system. (10)

42 What you find inside the pharynx of the Nereid worm. (4)

Down

1 In marine worms, the gonads form on these walls found between each metamere. (6)

3 This is the upper most part of a marine polychaete worm's parapodia. (10)

4 The repeating units in an annelid have this type of homology. (6)

5 The number of pairs of testes in an earthworm. (3)

7 Each of the repeating segments in an annelid is referred to as one of these. (8)

9 The trochophore larva found in annelids is also found in this phylum. (8)

12 These special glands help earthworms handle the extra calcium found in the food they ingest. (11)

13 The symbol for the metal found at the center of a polychaetes respiratory pigment. (2)

14 In addition to the fluid from the metanephridia, coelomic fluid in annelids also reaches the outside through these. (12)

16 This opening is found on the annelid's pygidium. (4)

17 Relative to the nerve cord, you'll find the ventral blood vessel in an earthworm here. (5)

18 The chlorogogue (chlorogogen) tissue is involved in this type metabolism. (8)

19 Deposit feeders eat this. (9)

21 Why there are three parts of an annelid that aren't segments can be found in this larval stage. (11)

23 Marine worms, polychaetes, and earthworms, oligochaetes, have these but leeches, hirudinea, don't. (5)

24 The extra rings of a leech. (6)

26 In annelids, structures made from this embryonic tissue are not metamerically arranged. (8)

27 The symbol for the metal found at the center of an oligochaetes respiratory pigment. (2)

28 The circulatory system of an annelid is this type. (6)

33 The specialized axons in the earthworm nervous system that help them escape into their burrows. (5)

34 Where you would find the subesophageal ganglion relative to the esophagus in a worm. (5)

35 The number of hearts in an earthworm has. (4)

Self Test - Annelida

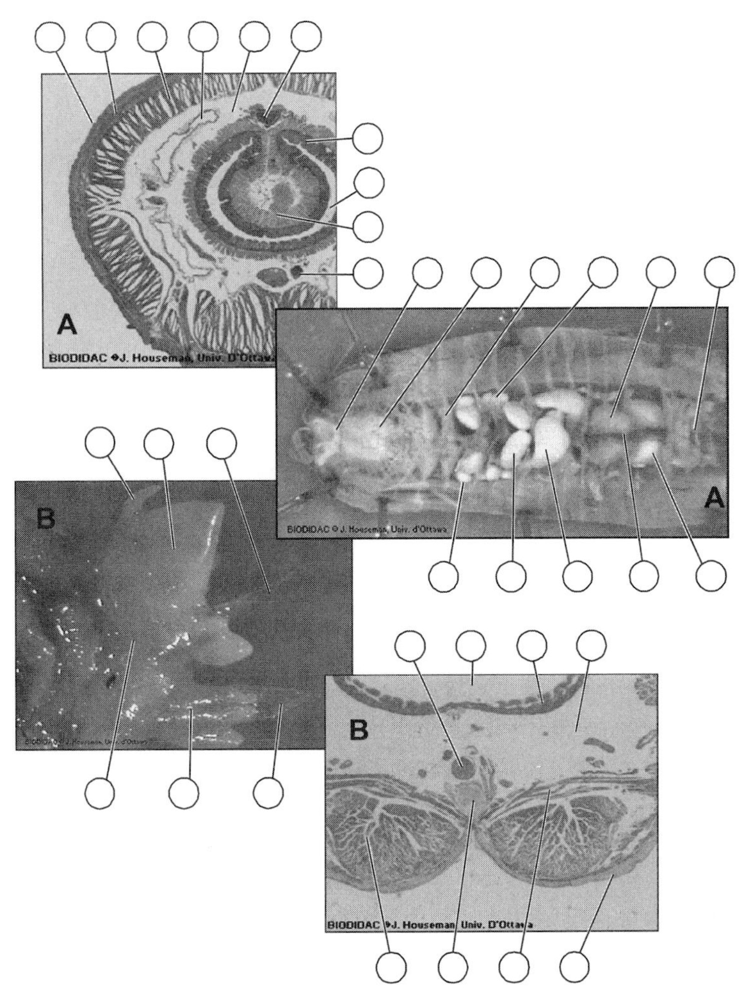

Use the following labels to identify the photographs. You may have to use a label more than once, and some labels may not be appropriate for the photographs. Answers are available on the *Digital Zoology* web site at www.mhhe.com/digitalzoology. Be sure to try the interactive Drag-and-Drop quizzes that are available on the *Digital Zoology* CD-ROM. A color version of this Self Test is available in the Adobe Acrobat version of the Student Workbook in the workbook folder on the *Digital Zoology* CD-ROM.

Specimens

A- Earthworm - *Lumbricus*

B- Nereid worm - *Nereis*

Labels

1) Brain
2) Circular muscle
3) Coelomic space
4) Crop
5) Dorsal cirrus
6) Dorsal vessel
7) Epidermis
8) First seminal receptacle
9) Gizzard
10) Gut lumen
11) Gut wall
12) Intestine
13) Longitudinal muscle
14) Metanephridium
15) Nerve cord
16) Neuropodium
17) Notopodium
18) Oblique muscle
19) Parapodium
20) Pharynx
21) Second seminal receptacle
22) Second seminal vesicle
23) Septal wall
24) Setal hairs
25) Third seminal vesicle
26) Typhlosole
27) Ventral vessel

MOLLUSCA

Inside *Digital Zoology*

As you explore the molluscs on the *Digital Zoology* CD-ROM, don't miss these learning tools:

Photos of dissected specimens of the preserved clam, snail, and squid.

Drag-and-drop quizzes on the key structures of the clam, snail, and squid.

Interactive cladogram showing the major events that gave rise to the eight mollusc classes. A summary of the key characteristics of each class are combined with an interactive glossary of terms.

Defining Differences

The characteristics described in the following sections appear for the first time in the molluscs, and they define the phylum. Others are important for understanding how animals in the phylum function. Molluscs have undergone adaptive radiation creating a bewildering array of animals that appear to have little or no resemblance to each other. It was once common practice in zoology courses to use a hypothetical mollusc that had the basic mollusc features to learn about animals in the phylum. It's unfortunate that it was called H.A.M. — hypothetical ancestral mollusc — because that gave the impression that this was what the ancestral mollusc might have looked like. It certainly didn't, and yet the exercise is a good one for understanding the phylum. So we'll rename H.A.M. — hypothetical aggregate mollusc — and still use it to discuss the different mollusc characteristics. You'll want to watch for examples of these in the various mollusc specimens that you will be examining.

- Radula,
- shell and a mantle,
- ctenidia,
- ciliated visceral mass,
- muscular foot, and an
- open circulatory system.

Radula

If you've found an animal with a **radula**, you've found a mollusc. The radula gave molluscs a way of feeding completely different from how any other animal fed. If you think about the different feeding strategies that ancestors to various animal phyla had, you'll see why. Particulate suspension feeders sat on the substrate or floated in the ancient oceans to capture their food — sponges, echinoderms and chordates. Flatworms glided across the ocean bottoms feeding on anything they could pick up with their fleshy muscular pharynx. That made platyhelminths one of the first to feed on the organic material that rained down and collected on the ocean bottom. Annelids tapped into a new unused food source by burrowing and consuming the substrate across which the flatworms were gliding. While all this was going on, predatory cnidarians were feeding on other animals in the zooplankton.

What about the organic matter "stuck" to the substrate that a flatworm couldn't get with only a fleshy pharynx? The solution was to scrape it off, and molluscs used their radula to do that. It's a simplistic view of things but for any animals that can exploit an unused food source there is the potential to diversify into new species. For molluscs, the radula, combined with the other traits that you will be looking at, allowed this phylum to undergo the extensive **adaptive radiation** that has made molluscs the second largest animal phylum. (Third, if there are as many nematodes as biologists think.)

Most textbooks compare the radula to a file, scraper, or rasp in a wood worker's tool box. (If you're more the power tool type — a belt sander.) It's a good analogy; the radula is a ribbon of teeth supported underneath by the tonguelike **odontophore**. As the radula is extended and retracted, it scrapes away at the substrate. If you rubbed a file on a piece of wood you would generate wood shavings or sawdust. It is the same with a radula, which creates particulate food. The usual solution for getting particulate food into the digestive tract is to trap it in **mucous**, then use **cilia** to propel the food-laden string of mucous into the mouth. As you might have guessed, that's what molluscs do. As the radula grinds away at the substrate the teeth on the end are worn down. They're replaced by new teeth added at the back of the radula, which as it grows, moves forward on the surface of the odontophore.

 *How does a mollusc separate the organic food from the inorganic substrate that it ingests?*

Shell and mantle

The shell originated as a defensive strategy by early molluscs to protect themselves using calcareous **spicules**, or spines, embedded in their outer epidermis. Spicules can still be found along with edges of some chiton shells and in some of the unshelled molluscs. Over time, the spicules became larger, forming plates that later fused with each other to create a solid shell covering the entire dorsal surface of the animal. The specialized epidermis and its gland cells that produced the shell are the **mantle. Retractor muscles** attached to inner surface of the shell extended into the foot on the ventral side and when these muscles contracted the shell was pulled down against the substrate. The mollusc hid inside waiting for whatever danger lurked outside to go away.

 *What are the three layers of the mollusc shell and of what is each made?*

It was a good defensive strategy but while it solved one problem, it created another. The new shell decreased the available surface area for gas exchange and any other functions that depended on diffusion. The problem was solved by extending the shell over the edge of the body, creating an open space underneath it — the **mantle cavity**. Just as the mantle on a fireplace hangs over the cavity of the fireplace, that's what the mollusc's mantle does, and the increased surface area for gas exchange occurs as gills that developed inside the cavity.

Ctenidia

Mollusc gills are called ctenidia, and when the cilia that cover them beat, water is pulled into the mantle cavity, across the surface of the ctenidia, and out the cavity. The ctenidia are hollow and the spaces inside of them are part of the hemocoel. As blood flows through ctenidia, it is oxygenated by the water passing by the outer surface.

The path that water takes through the mantle cavity is unidirectional. Once the water is past the ctenidia and on its way out, it goes by the openings of the **metanephridia**, gonad, and anus picking up gametes, if it's the reproductive season; metabolic wastes; and any undigested food from the alimentary tract, carrying it away from the mollusc.

Ciliated visceral mass

As important as cilia are in how the ctenidia work, they are equally important in most of the internal systems of a mollusc and why we make the point about a **ciliated visceral mass**. Cilia will propel the food through the digestive tract, and cilia on the metanephridia help propel the coelomic fluid into the excretory apparatus. Almost anything a mollusc does is done by using cilia, with one notable exception — how it moves.

Muscular foot

Although cilia initially played a role in movement, and may still found on the bottoms of some mollusc feet, muscles in the foot have become more important for how molluscs move; it's the reason the muscular foot is a molluscan trait.

There are four things that we need to keep track of to be able understand how the mollusc foot works, two of these are muscles and the other two are not muscles. One of two sets of muscles extend from the shell into the foot along the dorsal ventral axis of the body — the **dorsoventral muscles**. The other set of muscles run at right angles to the dorsoventral muscles from one side of the foot to the other — the **transverse muscles**. The two sets of muscles work in combination with the blood in the **hemocoel**, the third part, to form a **hydrostatic skeleton**. Think of the foot as a fluid-filled box. (When we looked at a hydrostatic skeleton in cnidarians we had sheets of muscles and fluid-filled balloon. This is different.) If we contract the dorsoventral muscles, the box could increase in either length, width, or both. But when the dorsoventral muscles contract, the transverse muscles contract at the same time. The result is that the box can only increase in length.

At the anterior end of the foot, contraction of the muscles extends the foot forward, and its tip contacts the substrate farther in front of the animal. Because the dorsoventral muscles have contracted, the part of the foot just behind where the most anterior is now touching, is raised off the substrate. Here's where the fourth element, **mucous**, comes in. The edges of the raised space underneath the contracted dorsoventral muscles are sealed to the substrate by mucous, like a small suction cup. As the dorsoventral muscles at the back of the suction cup contract, lifting that end, the dorsoventral muscles at the front relax. Because the space is sealed by mucous as the back end is lifted, the front end is sucked down, stretching the previously contracted dorsoventral muscles. A wave of contractions moves toward the back of the animal as the animal moves forward, a **retrograde wave**.

In some molluscs, each side of the foot is capable of separate waves of contraction, and depending on how fast the contractions move down either side, the animal turns and changes directions. It's like the treads on an armored tank, and with the way a mollusc protects itself using the shell, the analogy is even more appropriate.

Open circulatory system

In most zoology courses the molluscs are the first animals you'll encounter with an **open circulatory system**. In a **closed circulatory system,** exchange between the circulating blood and various tissues and organs occurs across **capillaries**. They're missing in an open circulatory system. Instead, blood pools in spaces or cavities, the **hemocoel**, and bathes the organs and tissues.

The hemocoel is the main body cavity of a mollusc and an important part of how molluscs move. There is a true coelom, but it is small, surrounding the heart. That's how it gets its name, the **pericardial cavity**. Gonads are found on the wall of the pericardial cavity, and an open funnel-shaped **metanephridia** filters the coelomic fluid in the cavity to remove metabolic wastes.

 *How do metabolic wastes get from blood in the hemocoel to the fluid in the pericardial cavity that the metanephridia filters?*

A Closer Look Inside *Digital Zoology* - Molluscs

Bivalvia: Freshwater clam

Clams and their bivalve relatives are one of only a few mollusc classes found in both freshwater and marine environments. In both places they're specialized as filter feeders, and like most filter feeders they're essentially sessile. To accommodate their way of life, bivalves have made some major modifications to the mollusc body plan. The most obvious is that the body is compressed laterally with the mantle folded down over the sides in two lobelike sheets. The shell matches the shape of the mantle's two parts. The shell has two valves that cover each side of the animal and the sides are connected by an elastic **hinge ligament** across the top. The two valves give the class their name, the bivalvia. But, be careful. These are **univalve** molluscs because the original single valve is only folded! Another modification for the filter-feeding lifestyle has resulted in the loss of the radula and cephalization.

How does a clam burrow in the sediment?

If you take a look at the **cladogram** for the molluscs in *Digital Zoology* or your textbook, you'll see that bivalves are more advanced than cephalopods, which include squids and octopods. Most students have trouble understanding how something like a highly intelligent octopus is considered more primitive than a sessile filter-feeding clam! **Cladistics** looks for shared characters and how they are modified from the ancestral form. If the character disappears completely, that's considered advanced. The things we think make the squids and octopods superior animals aren't characteristics that define the phylum. A radula defines the molluscs, and cephalopods still have their radula, and bivalves don't. It's one of a variety of reasons why the two groups are placed where they are on the cladogram.

The oldest part of the clam's shell is the **umbo,** and as the animal grows, the edge of the mantle grows out from the umbo. A clam keeps its shell closed by the contraction of large anterior and posterior **shell adductor muscles.** When the two muscles relax, the elasticity of the hinge opens the shell. It's these adductor muscles that let you know whether you should eat a steamed clam or mussel. Steaming kills the clam, and in death, muscles that hold the shell closed, relax and the shell pops open. If the shell didn't pop open when you cooked the clam, it was dead before it made it to the steamer.

✍ *What causes the ridged rings on the surface of the clam's shell?*

In the lab, once you've taken the shell off your clam you'll see the mother-of-pearl layer, or **nacre**, on the inside of the shell, along with the **pallial line** where the mantle connects and the **muscle scars** where the muscles attach. Fold the mantle back into place over the foot and ctenidia so that you can see how the left and right sides form the **incurrent siphon** and **excurrent siphon**. Freshwater clams have large ctenidia, which might lead you to believe that the animals need a lot of oxygen for something. However, they don't. The ctenidia are primarily involved in feeding. **Cilia** on the surface of the ctenidia pull water into the mantle cavity and filter out particulate food, passed forward along the edge of the ctenidia, to the **labial palps** and into the **mouth**.

✍ *Describe the route that water follows as it passes through a clam starting from the incurrent siphon and finishing with the excurrent siphon.*

The lateral compression of the bivalve body not only changes the shape of the shell but also the organs of the visceral mass. Instead of sitting above the muscular foot, the visceral mass is partially buried in the foot. The only part still visible above the foot is the **pericardial cavity** on the dorsal side. Inside it you'll see the **heart** that wraps around the **intestine** that passes through the cavity. If you are careful you'll also see the **atrium** to the heart. If you can't see it in your specimen, take a look in *Digital Zoology*. The atrium is important, and if you answered the question on how a **metanephridia** filters metabolic wastes, you'll know why. If you didn't answer that question earlier, now would be a good time.

To see the other organ systems in the foot, you'll have to patiently carve slices off the foot and slowly work your way deeper into the foot. As you do the intestines, **stomach**, **digestive glands**, and **gonads** will appear surrounded by the spongy musculature of the foot. Those spaces that create the sponginess are part of the **hemocoel**. If you have trouble seeing any of these, there are pictures in *Digital Zoology*.

Gastropoda: The snail

The gastropod trait of **torsion** makes the snail one of the more difficult invertebrate dissections. The **visceral mass** on top of the **muscular foot** has been rotated 180 degrees and the **mantle cavity** is now in front, sitting above the head. The opening of the mantle cavity points forward instead of toward the back, as it originally did. Another result of torsion is that structures caught on the inside of the bend are lost.

Snails often only have one of what would have been paired organs. These molluscs are **asymmetric** rather than bilaterally symmetric.

What are the presumed advantages of torsion, and for what stage in the gastropod life cycle was it an advantage?

To make things more complicated, snails are **monecious,** and to save space inside the cramped shell the two reproductive systems share some of same ducts. If that isn't confusing enough, the visceral mass is coiled inside a spiral shell! Be careful, though. The spiral shell is not a direct consequence of torsion. It's a solution for packaging the visceral mass. As the snail grows, the size of the visceral mass on top grows as well. It's wound up into the shell.

The snail's shell is an ideal hiding place for protection against predators, but it also protects against dissection. Snails are found in terrestrial, as well as freshwater and marine, environments. Terrestrial snails have lungs and breathe by raising and lowering the floor of the mantle cavity. This pumps air in and out of the **pneustome** located in the **collar** between the shell and foot. The successful invasion of land accounts for the success of the class.

Once you have your snail out of its shell, you'll have to start unwinding the viscera. Because of torsion, functional systems in the animal will wind their way up to the tip of the whorl and back down. The best example of this is the digestive system, which starts with the **buccal mass** under the mantle cavity, followed by the **esophagus** and the **stomach,** located in the first part of the whorl. The **intestine** leaves the whorl, passing by the first parts of the digestive tract on its way the anal opening. The **digestive gland** extends up and to the tip of the whorl.

The reproductive tract does the same thing, and some parts of the reproductive tract are easier to see than others. Use the pictures in *Digital Zoology* to help you orient with some of the more obvious landmarks in your specimen. The **dart sac**, **mucous gland**, **albumin gland**, and **flagellum** are usually the easiest to see. Once you've found those you should be able to follow the various tubes to find the **penis**; combined **vas deferens** and **oviduct**; **hermaphrodite duct**; **seminal receptacle;** and if you were especially careful, the **ovotestis** buried in the digestive gland.

 *In snails, describe the events from insemination to oviposition.*

Cephalopoda: The squid

A cephalopod's **visceral mass** has been stretched along the dorsoventral axis above the foot, bringing the head and foot closer together on the ventral side. That's how they got their name, Cephalopoda (head, foot). The **mantle** surrounds the visceral mass, and ancestrally a hard **shell** surrounded all of this to form an elongate cone-shaped shell with the head and foot poking out the open end. It was easier to point the tip of the shell in the direction that the animal was moving. Cephalopods swim with their dorsal surface pointing forward, rather than up.

With a larger visceral mass on top of a muscular foot, cephalopods were faced with the same problem as gastropods had for finding a way to make a more compact body. They used the same solution, although they wound their shells in a different way.

The fossil record includes cephalopods with unwound, wound, and even partially wound shells, and the diversity of fully wound ammonite fossils is good evidence of the successful body plan. At one time these cephalopod predators ruled the ancient oceans, but now only *Nautilus* remains to give a hint to what these ancient animals looked like.

✍ *How does the nautalid shell differ from that of a gastropod?*

Shelled cephalopods disappeared at about the same time as jawed predatory fishes appeared. It's possible that the two events are related. The mollusc adaptation to these more active and agile newcomers was another modification to the mollusc body plan to become just as active and agile — the result are the cephalopods we see today.

To live this new active and predatory life there have been extensive changes in how a cephalopod, like the squid, functions compared to other molluscs. Watch for these changes as you work with your squid in the lab or as you look at *Digital Zoology*.

One of the most obvious changes is the loss of the shell, or if it's still present a major reduction in its size. Externally you can't see a shell in your squid until you open the mantle and see the **pen** inside. Without a shell, the **mantle** is much more flexible, and its musculature is used to pump water in and out of the mantle cavity and across the gills. In other molluscs, cilia were responsible for this but they are completely missing on the gills of cephalopod. Take a look at the edge of the mantle on your squid, and you'll see the **mantle cartilages** that work like a lock and key to hold the mantle edge to the body. The mantle expands, pulling in around its edge. When it contracts, water is forced out the **funnel**. This movement of water doesn't only aerate the gills, it jet propels the animal. It's a new form of locomotion.

This enhanced water flow across the gills supplies levels of oxygen high enough to meet the metabolic demands of being an active predator, even though its flow is **concurrent** with the blood inside the gills. An open circulatory system with a hemocoel is not the most efficient way to move blood. Another cephalopod improvement is a **closed circulatory system** with capillaries supplying blood to the organs and tissues. To increase the efficiency of the system, a pair of **branchial hearts** push the blood across the gills to the systemic heart, which then pumps it out to the rest of the body. The branchial hearts are next to the **metanephridia** to remove metabolic wastes.

✍ *What route does blood follow as it moves through the squid? Start at the systemic heart.*

Being an active predator requires an advanced sensory and nervous system to locate and trap prey and to locate a mate and be able to differentiate that mate from a potential meal, or for that matter becoming one. Components of the system include a well-developed **brain** enclosed in a **cartilaginous brain case** and eyes analogous to our own. Cephalopods use **elastic capsule chromatophores** to change colors to signal mates or as camouflage. There are some cephalopods that can produce light, and their blinks and flashes are a communications tool.

Even the digestive tract has been extensively modified to fuel the metabolic demands. The **muscular foot** now forms the eight **arms** and paired **tentacles** used to capture prey using the suction cuplike **suckers**. There is still a **radula**, but a parrotlike **beak** is at the center of the ring of tentacles and arms, and it rips at food, which is swallowed and passed back into a digestive system. A **digestive caecum**, which acts like a crop as well, is the site of nutrient absorption. Other parts of the digestive system include the **stomach** and **digestive gland** (often called a liver) which produces digestive enzymes, and a small pouch on the digestive gland, the **pancreas**, that also produces digestive enzymes. The **esophagus** is buried inside the digestive gland and after the digestive caecum the digestive tract bends back on itself as the intestine leads towards the anus. Undigested food is dumped into the excurrent flow of the mantle cavity.

 *How do cephalopods solve the problem of getting sperm from the male to the genital pore of the female, hidden inside the mantle cavity?*

Cross-References to Other McGraw-Hill Zoology Titles

Integrated Principles of Zoology, 11th edition. C.P. Hickman, L.S. Roberts & A. Larson. Chapter 16.

Animal Diversity, 2nd edition. C.P. Hickman, L.S. Roberts & A. Larson. Chapter 9.

Zoology, 5th edition. S.A. Miller & J.P. Harley. Chapter 12.

Biology of the Invertebrates, 4th edition. J. Pechenik. Chapter 12.

Laboratory Studies in Integrated Principles of Zoology, 10th edition. C.P. Hickman, F. Hickman & L. Kats. Chapter 11.

General Zoology Laboratory Guide, 13th edition. C. Lytle. Chapter 11.

Structures Checklist

Here are some of the structures that you should be able to easily find in *Digital Zoology* and the specimens that you will be looking at in your lab. After reading your lab handout, you might want to add more and, depending on the equipment available in your lab, you might see more. As you study the material, you might also want to make some notes on how some of these structures looked or include a drawing in your lab notes. (Structures indicated by * may be hard to see.)

Clam

External anatomy

- ☐ Anterior end
- ☐ Excurrent siphon
- ☐ Growth lines
- ☐ Hinge ligament
- ☐ Incurrent siphon
- ☐ Left valve
- ☐ Periostracum
- ☐ Posterior end
- ☐ Right valve
- ☐ Umbo
- ☐
- ☐
- ☐
- ☐

Internal anatomy

- ☐ Anterior adductor muscle and scar
- ☐ Anterior aorta
- ☐ Anterior foot retractor muscle and scar
- ☐ Atrium
- ☐ Crystalline style*
- ☐ Ctenidia
- ☐ Digestive gland
- ☐ Esophagus
- ☐ Food groove on gill lamella
- ☐ Foot
- ☐ Gonad
- ☐ Heart
- ☐ Hinge teeth
- ☐ Intestine in foot
- ☐ Intestine in pericardial cavity
- ☐ Labial palp
- ☐ Mantle
- ☐ Metanephridium
- ☐ Mouth*
- ☐ Pallial line
- ☐ Pericardial cavity
- ☐ Posterior adductor muscle and scar
- ☐ Posterior aorta
- ☐ Posterior foot retractor muscle and scar
- ☐ Stomach
- ☐ Suprabranchial chamber
- ☐
- ☐
- ☐
- ☐
- ☐
- ☐

Additional structures

- ☐ Glochidia
- ☐
- ☐
- ☐
- ☐
- ☐
- ☐

Snail

External anatomy

- ☐ Anus*
- ☐ Coiled visceral mass
- ☐ Collar
- ☐ Eyes on second tentacle*
- ☐ Foot
- ☐ Genital pore*
- ☐ Growth lines
- ☐ Head
- ☐ Mantle
- ☐ Mouth
- ☐ Pneustome
- ☐ Shell aperture
- ☐ Shell apex
- ☐ Tentacles
- ☐ Whorl
- ☐
- ☐
- ☐
- ☐
- ☐

Internal anatomy

- ☐ Albumin gland
- ☐ Buccal mass
- ☐ Crop
- ☐ Dart sac
- ☐ Digestive glands
- ☐ Esophagus
- ☐ Flagellum
- ☐ Hermaphrodite duct
- ☐ Intestine
- ☐ Lung
- ☐ Mantle cavity
- ☐ Mucous gland
- ☐ Oviduct
- ☐ Ovitestis
- ☐ Penis
- ☐ Salivary glands
- ☐ Seminal duct
- ☐ Seminal receptacle
- ☐ Sperm duct (vas deferens)
- ☐ Stomach*
- ☐
- ☐
- ☐
- ☐
- ☐
- ☐

Additional structures

- ☐ Shell columella
- ☐
- ☐
- ☐
- ☐

Squid

External anatomy

- ☐ Anterior surface
- ☐ Arms
- ☐ Beak
- ☐ Buccal membrane
- ☐ Collar
- ☐ Dorsal surface
- ☐ Eyes
- ☐ Head
- ☐ Lateral fins
- ☐ Mantle
- ☐ Mantle cartilages
- ☐ Mantle collar
- ☐ Posterior surface
- ☐ Suckers
- ☐ Tentacles
- ☐ Ventral surface
- ☐
- ☐
- ☐
- ☐
- ☐
- ☐

Internal anatomy

- ☐ Afferent branchial vein
- ☐ Anterior vena cava (Cephalic vein)
- ☐ Brain
- ☐ Branchial hearts
- ☐ Cephalic aorta
- ☐ Cephalic cartilage
- ☐ Digestive caecum
- ☐ Digestive gland (liver)
- ☐ Efferent branchial vein
- ☐ Esophagus*
- ☐ Funnel retractor muscles
- ☐ Gills
- ☐ Head retractor muscles
- ☐ Ink sac
- ☐ Intestine
- ☐ Mantle arteries
- ☐ Metanephridia (kidneys)
- ☐ Nidamental gland
- ☐ Ovaries
- ☐ Oviduct
- ☐ Oviducal opening
- ☐ Pancreas
- ☐ Pen
- ☐ Posterior aorta
- ☐ Posterior vena cava (lateral abdominal vein)
- ☐ Rectum
- ☐ Spermatophoric duct
- ☐ Spermatophoric sac (Needham's sac)
- ☐ Systemic hearts
- ☐ Testes
- ☐
- ☐
- ☐
- ☐
- ☐

Additional structures

- ☐ Hectocotylized arm in male
- ☐
- ☐
- ☐
- ☐
- ☐

Crossword Puzzle – Mollusca

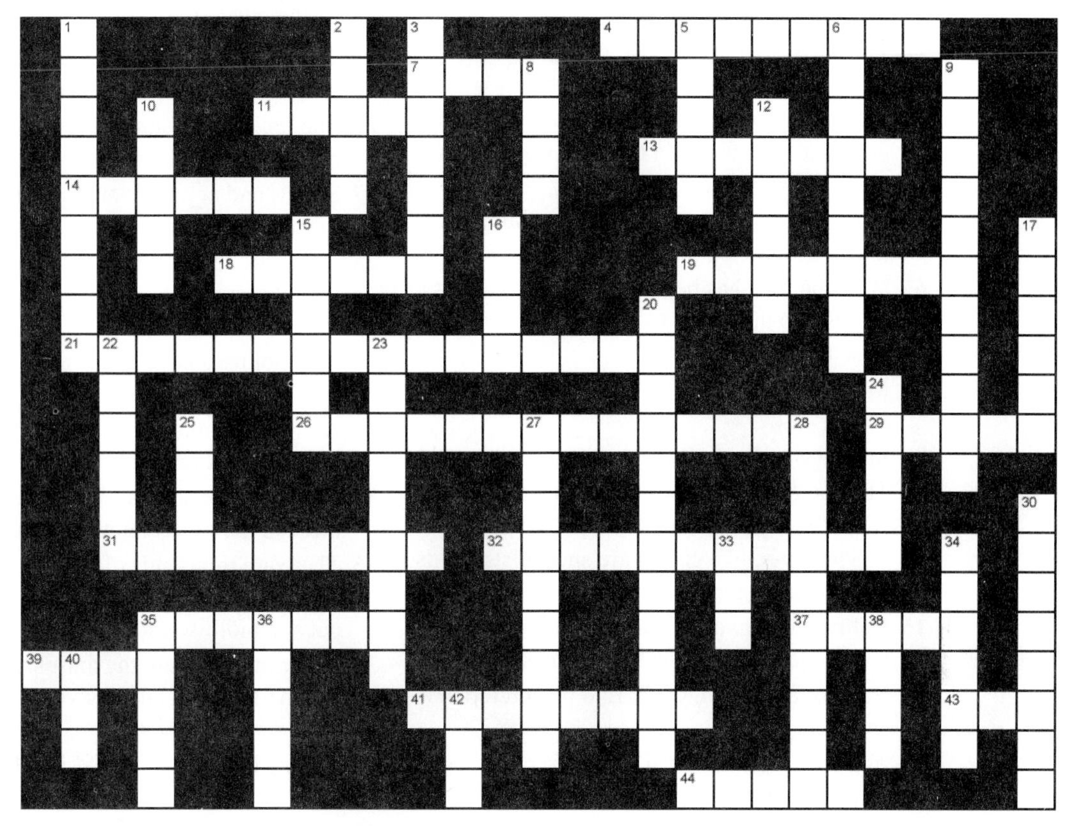

Puzzle solution. An interactive web based version of this puzzle, and its solution, are available on the *Digital Zoology* web site at www.mhhe.com/DigitalZoology/Students. With the interactive puzzle you can check to see if individual words or the whole puzzle is correct, and get hints for single letters.

Across

4 These are the larval stages of the freshwater clams. (9)

7 The oldest part of the bivalve clam shell. (4)

11 The blood passes through these just before entering the heart in a mollusc. (5)

13 One of the other larval stages common in molluscs. (7)

14 The edge of this tissue secretes the shell and creates the cavity that bears its name. (6)

18 When the shell adductor muscle contracts in a clam, this happens to the shell. (6)

19 Describes the molluscan gill. (8)

21 Digestive enzymes in the stomach of the clam are released from this spinning structure. (11,5)

26 This filters the fluid of the pericardial cavity in a mollusc. (14)

29 This part of the circulatory system is inside the pericardial cavity of molluscs. (5)

31 Instead of ancestral, the A in H.A.M. should be this. (9)

32 The radula in a mollusc sits on this cartilaginous tonguelike structure. (11)

35 The asymmetric body plan of a snail is a result of this. (7)

37 These structures are used to propel water across the mollusc ctenidia. (5)

39 The radula resembles this type of tool found in a toolbox. (4)

41 The more technical term for the mother-of-pearl layer of a mollusc shell is this layer. (8)

43 The shell of a clam is composed of this number of valves. (3)

44 True or false: The spiral shell of a mollusc is a direct result of torsion. (5)

Down

1 The inner shell layer in a mollusc. (9)

2 Describes each half of a clam's shell. (5)

3 In squids the gills are no longer covered in cilia now that these are used to pump water in and out of the mantle cavity. (7)

5 Relative to the other layers of a mollusc shell, the position of the periostracum. (5)

6 Once the food has been sorted in a mollusc, it passes into this gland to be biochemically broken down. (9)

8 Molluscs have this type of a circulatory system. (4)

9 In molluscs, the true coelom is this cavity. (11)

10 This ligament makes the clam shell pop open when the adductor muscles relax. (5)

12 The bivalves, like clams, are specialists at this type of feeding. (6)

15 Unlike the flatworms that came before them, the molluscs added this to their mesoderm. (6)

16 Because of torsion, you'll find this on top of a snail's head. (4)

17 There's no ciliated larval stages in cephalopods; instead, miniature animals that resemble the adults hatch from the egg. It's referred to as this type of development. (6)

20 Conchiolin is a protein embedded in this layer of the mollusc shell and helps protect the shell. (12)

22 The unique molluscan feeding structure. (6)

23 In clams, the heart wraps around this as it passes through the pericardial cavity. (9)

24 To improve circulation, predatory molluscs, such as the squid, have this many hearts. (5)

25 The brain of most molluscs resembles this because it surrounds the esophagus. (4)

27 Squids and octopods are molluscs adapted to this type of feeding lifestyle. (9)

28 Snails are referred to as this because both sexual organ systems are found in each animal. (10)

30 The molluscs have an open circulatory system and blood pools here. (8)

33 All that remains of the mollusc shell in a squid is a short rod referred to as this. (3)

34 Highly active squids have more than one of these in their circulatory system. (6)

35 There are thousands of these on the surface of the radula, and they can scrape, pierce, tear, or cut at a mollusc's food. (5)

36 You can tell where the muscles of the clam attach to its shell because you'll see these on the shell surface. (5)

38 Terrestrial snails breath using this instead of gills. (4)

40 Secretions from this gland act as decoy allowing cephalopods to make their escape. (3)

42 The material that flows through a pneustome. (3)

Self Test - Mollusca

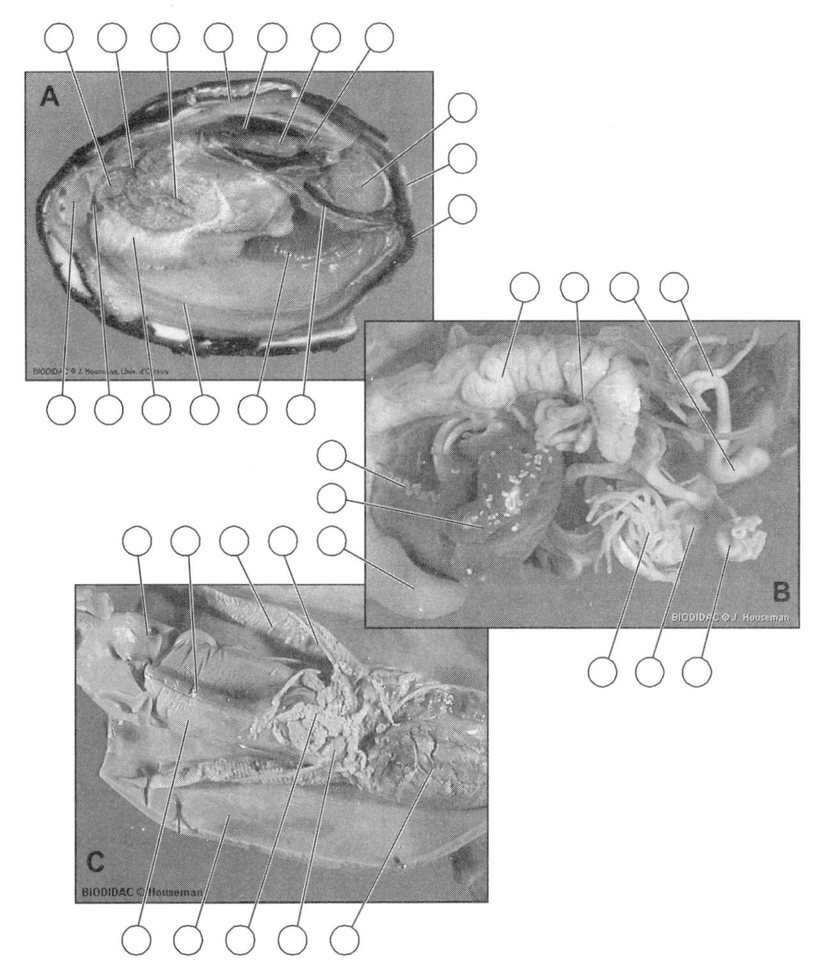

Use the following labels to identify the photographs. You may have to use a label more than once, and some labels may not be appropriate for the photographs. Answers are available on the *Digital Zoology* web site at www.mhhe.com/digitalzoology. Be sure to try the interactive Drag-and-Drop quizzes that are available on the *Digital Zoology* CD-ROM. A color version of this Self Test is available in the Adobe Acrobat version of the Student Workbook in the workbook folder on the *Digital Zoology* CD-ROM.

Specimens

A- Clam
B- Snail
C- Squid

Labels

1) Afferent branchial vein
2) Albumen gland
3) Anterior shell adductor
4) Anterior vena cava
5) Branchial heart
6) Ctenidia
7) Dart sac
8) Digestive caecum
9) Digestive gland
10) Excurrent siphon
11) Flagellum
12) Foot
13) Heart
14) Hermaphrodite duct
15) Hinge
16) Incurrent siphon
17) Intestine
18) Mantle
19) Mantle cartilage
20) Mouth
21) Mucous gland
22) Oviduct
23) Pancreas
24) Penis
25) Pericardial cavity
26) Posterior shell adductor
27) Seminal receptacle
28) Stomach
29) Suprabranchial chamber
30) Vas deferens

ARTHROPODA

Inside *Digital Zoology*

As you explore the arthropods on the *Digital Zoology* CD-ROM, don't miss these learning tools:

- **P**hotos of intact and dissected specimens of the crustacean crab (*Cancer*) and crayfish (*Cambarus*) and the uniramian grasshopper (*Romalea*) and cockroach (*Periplaneta*).
- **V**ideo clips of the circulatory system of the cockroach showing the heart beating and blood flowing through the veins of the wing.
- **D**rag-and-drop quizzes on the key internal and external structures of the crab, crayfish, cockroach, and the external anatomy of the grasshopper.
- **I**nteractive cladograms for the three arthropod subphyla: Chelicerata, Uniramia, and Crustacea. A summary of the key characteristics of each class are combined with an interactive glossary of terms.

Defining Differences

Some of the differences described in the following sections are features that arise for the first time in the Arthropoda. They define the phylum. Others are important for how animals in the phylum function. Arthropods are the most abundant animal phylum on earth, and they make that claim because of the tremendous numbers of insects included in the subphylum Uniramia. There's a lot to look at, and the features we've list apply to all arthropods, although in many of the subphyla they have been extensively altered and may even have disappeared. Whichever is the case, you'll want to watch for examples of these in the various arthropod specimens that you will be examining.

- Tagmatization of the
- chitinous exoskeleton, growth by
- molting,
- muscle arranged in bands,
- jointed appendages, and
- compound eyes.

Tagmatization

An annelid's body is made up of a series of repeating units sequentially arranged down the length of the animal. **Metamerisation** describes the process that resulted in the segments, **metameres**, each of which is identical to the one on either side of it. Arthropods took metamerisation a step further. **Tagmatization** grouped adjacent metameres into larger functional units, **tagma**, responsible for performing specialized tasks. In addition to how the segments were organized into tagma, another important difference between the two phyla is that arthropods have paired appendages on each their metameres.

A word of warning! This sequence of events assumes that arthropods were the logical evolutionary step that followed annelids. There is growing evidence that this may not be the case. This was discussed earlier in the workbook when we looked at the first animals with a cuticle, the nematodes. If the **Ecdysozoa** is a valid **monophyletic** taxon, then this annelid-to-arthropod transition is inappropriate. While we wait for a resolution to this taxonomic conundrum, the transition from a metameric body plan to one based on tagma is probably correct. The question is whether an annelid is the right metameric ancestor.

To understand the impact tagmatization has on how an animal functions, let's take a look at the ancestral and modified body plans of two different arthropod subphyla: Crustacea and Uniramia.

Primitive crustaceans had only two tagma, a **head** and a **trunk**. The head was formed from five segments and their five pairs of appendages. This included two pairs of sensory appendages: **antennae and antennules,** and three pairs of feeding appendages: **mandibles**, **first maxillae,** and **second maxillae** used for grinding, cutting, and manipulating food before it was ingested. Each of the trunk appendages had three responsibilities: locomotion, respiration, and food gathering. As this primitive crustacean paddled along using its trunk appendages, it trapped particulate food using **setal hairs** on the surface and passed it forward to the mouth. While all this was happening, the large surface area of the legs was used for gas exchange. Compare that to the lobster or crayfish. The head tagma has been replaced with a **cephalothorax.** Some of the trunk appendages have been recruited to help with feeding and **maxillipeds** are part of the "cephalo" part of the cephalothorax. The "thorax" part, more correctly the **pereon** with its **pereopods**, is locomotary and houses the main organ systems. The rest of the trunk, **pleon** with **pleopods**, is a muscular tail used to escape from predators and in the females for brooding their eggs.

In the Uniramia centipedes, the head tagma is a fusion of six segments followed by the trunk tagma with a pair of walking legs on each of the segments. The uniramian head hasn't changed much between centipedes and insects. It processes sensory information from the **antennae** and **compound eyes** and collects and manipulates food before it's ingested. The biggest difference can be seen in changes to the trunk, which in insects is now the **thorax** and **abdomen.** This second insect tagma, the thorax, is specialized for locomotion, and the only thing you'll find inside it are muscles that move the wings and legs. The organ systems pass their ducts and tubes through the thorax on their way to the third tagma, the abdomen, which houses the main organ systems. The abdomen's role is the day-to-day functioning of the animal, and it contains the reproductive system.

✍ *We've described the ancestral tagma of the two of the arthropod subphyla; how are the tagma and appendages arranged in a Chelicerata?*

As the tagma take on these different roles, their shapes, and the shapes of the appendages attached to them, change. Arthropod cuticle is a "living plastic" and can be molded into almost anything you could think of. The result is a tremendous array of different cuticular shapes and forms, many of which are arthropod tools. For example, cuticular appendages can be delicate sensory antennae, shovels, scissors, grinding stones, walking legs, defensive pincers, and paddles for swimming. The possibilities are almost endless, and it's one of the reasons why arthropods are such a successful group.

Chitinous exoskeleton

Invertebrate cuticles are made from either **collagen** or **chitin.** Arthropods use chitin, a long chain carbohydrate polymer of modified glucose units, **N-acetyl-glucosomaine** to be more precise. The use as a structural fiber is not new with the arthropods. The cell walls in plants are strengthened with cellulose fibers, chains of glucose units. Like arthropods, fungi use chitin in their cells.

Arthropod cuticle has two important functions. The first is a combination of strength and support and the **exoskeleton** creates a suit of armor that protects the animal inside. The cuticle's other function, as a chemical barrier, may not be as obvious. The exoskeleton prevents things from diffusing out of the animal or unwanted substances from diffusing in. This is a particularly important role for terrestrial arthropods, which need to keep water inside.

The strength of the arthropod's cuticle is a mix of chitin and cuticular proteins. This mix is solidified in the **procuticle** by a combination of chemical cross-linking and/or embedded calcium salts. Chelicerates and uniramians use chemical cross-linking and crustaceans use calcium salts; it's why when you crack into your lobster claws or crab legs you hear that crusty, crunching sound.

The chemical barrier is created by the outer **epicuticle,** and in terrestrial arthropods it involves a complex mixture of proteins and waxes that waterproof the animal. Arthropods were the first animals to conquer the terrestrial environment, and they have the most complex epicuticles. Waterproofing the cuticle was an important event in the evolution of terrestrial arthropods.

 *Why don't crustaceans have a waterproof epicuticle? What's the consequence of not having one?*

By varying the depth of procuticle, and/or the degree of cross-linking or embedded salts, the cuticle can be thick and hard or thin and membranous. Hard plates, **sclerites**, are connected to each other by membranous areas, which allows for bending and movement of the arthropod's armor.

Molting

There's one problem with living in a suit of armor. As the animal grows it needs to discard its current shell for a new, larger suit of armor. That's what arthropods do, and the process is referred to as **ecdysis** or **molting**. Building a suit of armor was a major metabolic investment for an arthropod. There's more to molting than discarding the old cuticle. Arthropods recycle parts of their old cuticle to help build the new one. **Chitinases** breakdown the chitin and **proteinases** attack the proteins to regenerate N-acetyl-glucosamine and the amino acid subunits that help to build the new cuticle.

 *Why do the chitinases and proteinases breakdown only the old cuticle and not the new one?*

Muscle arranged in bands

A solid outer cuticle has affected the way that muscles are arranged in arthropods. Until this point in our discussions of different animals, the **hydrostatic skeletons** have been the most common type of skeleton. Sheets of muscles are arranged in opposite directions to each other, circular and longitudinal, and surround a fluid filled space. Even nematodes, with their cuticle, still had a way to make it flexible enough to be used as hydrostatic skeleton for movement. The arthropod's cuticle isn't like a nematode's; it's solid, making the use of a hydrostatic skeleton impossible. Sheets of muscles are replaced by bands of muscle embedded in solid cuticle on either side of a membranous piece of cuticle. As the muscle contracts and shortens, it moves one piece of cuticle relative to the other. This type of muscular arrangement, in combination with the tubular exoskeleton, creates the unique jointed limbs and feet that give the phylum its name, the Arthropoda – jointed foot animals.

Jointed appendages

In its simplest form, an arthropod appendage is a tubular shell of cuticle surrounding muscles. There is only one type of joint that can be used to make this bend, and the two tubes on either side of the joint come in contact with each other at two points, the hinges or pivot points. When one tube moves relative to the other it bends in only one direction; pivoting around the line between the two hinges. Having only one of these joints on an appendage isn't enough. Further along are more of these joints, each of which bends at a different angle. With each joint bending in its own direction the tip of the appendage can be positioned anywhere within its reach.

We started with the analogy that an insect's exoskeleton was like a suit of armor and joints in the arms and legs of a suit of armor work the same way that they do in arthropods. From ancient armor, to robotic arms on an automobile assembly line, to the arm on the space shuttle or space station, they all have the same design constraints as arthropod appendages.

There are biomechanical limits to how big structures can become with this type of tubular construction. As the appendage becomes larger, the wall of the tube has to become thicker to withstand forces generated by the larger muscles inside. This increasing thickness means increased weight. There's a point where the muscle inside the tube can't lift the weight of the appendage. The way around the problem is to place the muscles on the outside of a solid supporting skeletal rod, an endoskeleton—a strategy used by the chordates.

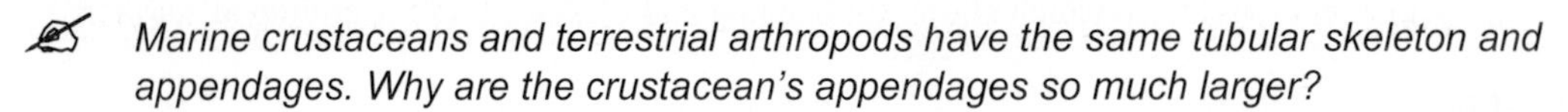

Marine crustaceans and terrestrial arthropods have the same tubular skeleton and appendages. Why are the crustacean's appendages so much larger?

Compound eyes

Arthropods have a unique visual system, the **compound eyes**. Or at least most of them do; it is missing in many of the chelicerates. Presumably, its absence is an advanced trait because compound eyes are found in primitive horseshoe crabs, also chelicerates.

The compound eye is made up of a series of optic units, the **ommatidia,** the fundamental units for creating the image. Each ommatidia is formed from a series of light-sensitive **retinular cells** optically isolated from each other by **pigment cells**. Different retinular cells are responsive to different wavelengths of light, and their responsiveness corresponds to three primary colors. The cells detect different intensities of color as they interpret the small part of the bigger picture that it samples. Each of these individual pieces of the picture is combined by the arthropod brain to create the final image.

Think of each of the ommatidia as a pixel on a computer screen or television. The retinular cells input a value for each of the red, green, or blue color depths of the pixel. You can see images and motion on the screen as the pixels change their colors. The arthropod eye works much like this, with one notable exception with insects. Insects don't see the color red at the upper end of the color spectrum, but do see ultraviolet at the lower end of the spectrum.

Although the compound eye is usually identified as an arthropod trait, arthropods also have **simple eyes** or **ocelli**. The simple eyes don't usually form images and are usually involved in detecting light intensity. A notable exception to this are the large modified simple eyes of hunting spiders. The loss of the image forming compound eye early in the evolution of the chelicerates left these spiders with only the simple eyes that could be modified for forming images.

✍ *Eyes are important in detecting and capturing prey. How do chelicerates do that without compound eyes?*

A Closer Look Inside *Digital Zoology* - Arthropoda

Crustacea: Crayfish and crab

One of the traditional lab exercises in zoology is to examine **serial homology** in the legs of a crayfish or lobster. (*Digital Zoology* also has a crab that you can do this with as well.) One of the differences between annelid and arthropod **metameres** is arthropod segments have a pair of appendages. In chelicerates and uniramians, they are **uniramous**; crustacean appendages are **biramous**. Each appendage has two branches, an outer **exopodite** and an inner **endopodite** attached to a single **protopodite** that anchors the limb on the body. The protopodite may have cuticular extensions on their inside surface, **endites**, or the outer surface, **exites**. Any, or all of these, can be modified in different ways for locomotion, feeding, or respiration. As you examine each of the appendages in your lab, or with *Digital Zoology*, you'll see an example of the "living plastic" of arthropod cuticle and how it can be molded into a variety of shapes. That's the reason in almost every zoology course you'll find students pulling the appendages of their specimen and lining them up on the bench – it's almost a right of passage.

Crustaceans have an **open circulatory system,** and their **heart** is located inside a **pericardial cavity** on the dorsal side of the animal. Crab and crayfish blood contains the respiratory pigment **hemocyanin,** and it's essential that all tissues receive oxygenated blood. Three major blood vessels supply blood to the anterior (**anterior aorta**), posterior (**posterior aorta**), and ventral surface (**sternal artery**) of the animal. Each vessel divides further into smaller vessels that supply blood to distinct organs or parts of the body. There are no capillaries in the system, and oxygenated blood is poured over the organs and tissues and then pools in the **hemocoel**. Here it continues to supply oxygen and nutrients, and it picks up metabolic wastes from the structures that it bathes. As the crustacean heart beats it pulls blood from the hemocoel through the **gills**, and out through the three major blood vessels. Gills in these large crustaceans are housed inside **branchial chambers. Gill bailers** help to keep water moving across the respiratory surfaces.

✍ *How do crabs and crayfish eliminate metabolic wastes?*

Crayfish and crabs have an unusual **gastric mill**, or **stomach,** as part of their digestive system. What makes it unusual are the three hard chitinous teeth inside. Students often wonder how there can be teeth so far down the digestive tract. The answer is in how the gut forms in arthropods and other animals. The gut forms from **ectodermal** invaginations from the anterior and posterior end of the embryo that join with the **endoderm** in the middle. This creates the **archenteron**, or primitive gut. An arthropod's ectoderm is lined with cuticle, and the ectodermal front and back parts of the gut have a cuticular lining. The gastric mill is part of the ectodermal lining. There is, however, a potential problem with having a cuticle lined gut, and it's related to the cuticle's role as a chemical barrier. Digestive enzymes can't be secreted across the cuticle, and nutrients can't be absorbed, either. That has to be done by the part of the gut formed from the endoderm. In these two animals, this is the job of the endodermal **digestive gland** and the **midgut,** which has no cuticular lining.

✍ *How does locomotion differ in the crab compared to crayfish or lobsters?*

You shouldn't be left with the impression that all crustaceans are the size of the two specimens in *Digital Zoology*. The vast majority of crustaceans are microscopic zooplankton of aquatic environments. These minute crustaceans occupy a critical position in marine food chains as the primary herbivores in that environment.

Uniramia: Cockroach and grasshopper

Over 85 percent of the animals on the planet, over a million species, are insects. They all look remarkably similar to each other. If it's an arthropod with three pairs of legs, it's an insect. You might be tempted to include wings in the definition but primitive insects were wingless making wings a **derived trait**, rather than an **ancestral trait** for the group. There are a number of reasons why insects are so successful. One of the more important was their being the first animals to successfully colonize land. They were able to do this because they found a way to waterproof their bodies and conserve water. One way they did this was through a structural modification to the cuticle; the other was a functional change that resulted in a unique excretory system – **Malpighian tubules**. More precisely, it should be referred to as the **Malpighian tubule hindgut complex** because **rectal pads** in the **hindgut** are responsible for absorbing the last of the water from the fecal material before it leaves the animal.

Insects have an extra **waxy layer** in their **epicuticle** that waterproofs the whole surface of the animal. It's combined with a complex mixture of cross-linked proteins, the **cement layer**, that hardens the surface against abrasion, another potential problem in the terrestrial environment.

Uric acid is the main metabolic waste in insects. It is not very toxic, and it packs more nitrogen per molecule than either **urea** or **ammonia**. Malpighian tubules are closed at one end and empty into the gut at the junction between the **midgut** and **hindgut**. The tubules are bathed in hemolymph and create an isotonic **primary urine**, a hemolymph ultrafiltrate of the hemolymph containing molecules small enough to pass through the membrane of the tubule cells: water, salts, low molecular weight nutrients, and dissolved uric acid. This is mixed with the undigested food as it leaves the midgut, and the hindgut absorbs water, salts, and nutrients contained in the mix of fecal material and urine and leaves the uric acid in the rectum. Depending on where an insect lives, and how limited water is, the hindgut absorbs the appropriate amount of water. Insects that live inside seeds for their whole life never have the opportunity to drink water, and their Malpighian tubule hindgut complex is so efficient at recovering water that metabolic water generated by metabolic process is sufficient for the insect to survive. The fecal material is a dry dust.

Insects have an **open circulatory system,** and their dorsal heart pumps blood toward the head. From there it percolates back through the large hemocoel bathing the organs and structures housed inside the cavity. Unlike crustacean blood, or the blood from most of the other animals, insect blood is not involved in oxygen transport. Insects have a **tracheal system** that supplies air directly to the tissues. **Spiracles** open in the side of the thorax and abdomen and are connected to **tracheal trunks** strengthened by rings of cuticle, **tenidia**, which prevent the trunks from collapsing. Their tracheal trunks connect to **trachea,** which branch throughout the insect's body. The large surface area created by the tracheal system is a potential site for water loss. That doesn't happen, however, because the trachea are covered with the waxy epicuticle. Gas exchange occurs across the smaller **trachioles,** which don't have this waterproof covering. The tracheole's microscopic fingers extend and wrap around the cells into which oxygen diffuses. With the blood no longer involved in oxygen transport, it functions more as a tissue culture medium supplying tissues with nutrients and materials they need and carrying away the metabolic wastes. Insect blood is high in protein, sugars, fats, and amino acids.

Cross-References to Other McGraw-Hill Zoology Titles

Integrated Principles of Zoology, 11th edition. C.P. Hickman, L.S. Roberts & A. Larson. Chapter 18-20.

Animal Diversity, 2nd edition. C.P. Hickman, L.S. Roberts & A. Larson. Chapter 11.

Zoology, 5th edition. S.A. Miller & J.P. Harley. Chapter 14 & 15.

Biology of the Invertebrates, 4th edition. J. Pechenik. Chapter 14.

Laboratory Studies in Integrated Principles of Zoology, 10th edition. C.P. Hickman, F. Hickman & L. Kats. Chapter 13-15.

General Zoology Laboratory Guide, 13th edition. C. Lytle. Chapter 13.

Structures Checklist

Here are some of the structures that you should be able to easily find in *Digital Zoology* and the specimens that you will be looking at in your lab. After reading your lab handout, you might want to add more and, depending on the equipment available in your lab, you might see more. As you study the material, you might also want to make some notes on how some of these structures looked or include a drawing in your lab notes. (Structures indicated by * may be hard to see.)

Crayfish

External anatomy

- ☐ Abdomen
- ☐ Antennae
- ☐ Antennules
- ☐ Anus
- ☐ Branchial chamber
- ☐ Carapace
- ☐ Cepahlothorax
- ☐ Cervical groove
- ☐ Cheliped
- ☐ Compund eye
- ☐ First maxilla
- ☐ First maxilliped
- ☐ Gonopods (Male swimmeret)
- ☐ Mandible
- ☐ Mouth
- ☐ Opening of antennal gland
- ☐ Rostrum
- ☐ Second maxilla
- ☐ Second maxilliped
- ☐ Sternum
- ☐ Swimmerets (Pleopods)
- ☐ Telson
- ☐ Uropods
- ☐ Walking legs
- ☐
- ☐
- ☐
- ☐
- ☐

Dissection

- ☐ Abdominal extensor muscle
- ☐ Abdominal flexor muscle
- ☐ Antennal artery
- ☐ Antennal gland
- ☐ Anterior gastric muscle
- ☐ Digestive gland
- ☐ Dorsal abdominal artery
- ☐ Esophagus
- ☐ Gastric mill pyloric region
- ☐ Gastric mill, cardiac region
- ☐ Gastric teeth
- ☐ Gill bailer (Scaphognathite)
- ☐ Gills, arthrobranchs
- ☐ Gills, podobranchs
- ☐ Heart
- ☐ Heart ostia
- ☐ Hepatic artery
- ☐ Intestine
- ☐ Mandibular muscle
- ☐ Nerve cord
- ☐ Ophthalmic artery
- ☐ Ovary
- ☐ Oviduct*
- ☐ Pericardial cavity
- ☐ Posterior gastric muscle
- ☐ Sperm ducts*
- ☐ Sternal artery
- ☐ Supraesophagial ganglion (Brain)*
- ☐ Testes
- ☐ Ventral abdominal artery
- ☐ Ventral thoracic artery
- ☐
- ☐
- ☐
- ☐

Additional structures

- ☐
- ☐
- ☐
- ☐
- ☐
- ☐
- ☐

Crab

External anatomy

- ☐ Abdomen
- ☐ Antennae
- ☐ Antennules
- ☐ Cephalothorax
- ☐ Cheliped
- ☐ Compound eye
- ☐ Exhalan opening to gill chamber
- ☐ First maxilla
- ☐ First maxilliped
- ☐ Flabella on maxillipeds
- ☐ Gill bailer (Scaphognathite)
- ☐ Inhalant opening to gill chamber
- ☐ Mandible
- ☐ Mouth
- ☐ Pleopods
- ☐ Second maxilla
- ☐ Second maxilliped
- ☐ Sub-branchial carapace
- ☐ Third maxilliped
- ☐ Thoracic sterna
- ☐ Walking legs
- ☐
- ☐
- ☐
- ☐

Dissection

- ☐ Digestive gland
- ☐ Gastric mill, cardiac region
- ☐ Gastric mill, pyloric region
- ☐ Anterior gastric muscle
- ☐ Posterior gastric muscle
- ☐ Pericardial cavity
- ☐ Heart
- ☐ Heart ostia
- ☐ Ovary
- ☐ Oviduct
- ☐ Testes
- ☐ Sperm duct
- ☐ Antennal glands*
- ☐ Gills
- ☐ Mandibular muscle
- ☐ Gill cleaner
- ☐
- ☐
- ☐

Additional structures

- ☐ Dimorphism of the male and female abdomen
- ☐
- ☐
- ☐

Grasshopper and Cockroach

External anatomy

- ☐ Abdomen
- ☐ Antennae
- ☐ Arolium
- ☐ Cerci (Cockroach only)
- ☐ Claw
- ☐ Compound eyes
- ☐ Coxa
- ☐ Femur
- ☐ Forewing
- ☐ Head
- ☐ Hindwing
- ☐ Hypopharynx
- ☐ Labial palps
- ☐ Labium
- ☐ Labrum
- ☐ Legs
- ☐ Mandible
- ☐ Maxilla
- ☐ Maxillary palps
- ☐ Mesothorax
- ☐ Metathorax
- ☐ Ovipositor
- ☐ Ovipositor
- ☐ Prothorax
- ☐ Simple eyes, ocelli
- ☐ Spiracles
- ☐ Sternites
- ☐ Tarsomeres
- ☐ Tarsus
- ☐ Tergites
- ☐ Tibia
- ☐ Trochanter
- ☐ Tympanum (Grasshopper only)
- ☐
- ☐
- ☐
- ☐
- ☐

Dissection

- ☐ Abdominal ganglia
- ☐ Digestive ceaca
- ☐ Esophagus
- ☐ Fat body
- ☐ Female accessory glands
- ☐ Flight muscles
- ☐ Heart
- ☐ Male accesory glands
- ☐ Malpighian tubules
- ☐ Midgut
- ☐ Nerve cord
- ☐ Ovary
- ☐ Oviduct
- ☐ Rectum
- ☐ Salivary glands
- ☐ Sperm duct*
- ☐ Stomach
- ☐ Subesophageal ganglion*
- ☐ Supraesophageal ganglion (Brain)*
- ☐ Testes
- ☐ Thoracic ganglia
- ☐ Trachea
- ☐ Tracheal trunks
- ☐
- ☐
- ☐
- ☐
- ☐

Additional structures

- ☐ Immature stages of the life cycle
- ☐ Ootheca of the cockroach
- ☐

Crossword Puzzle – Arthropoda

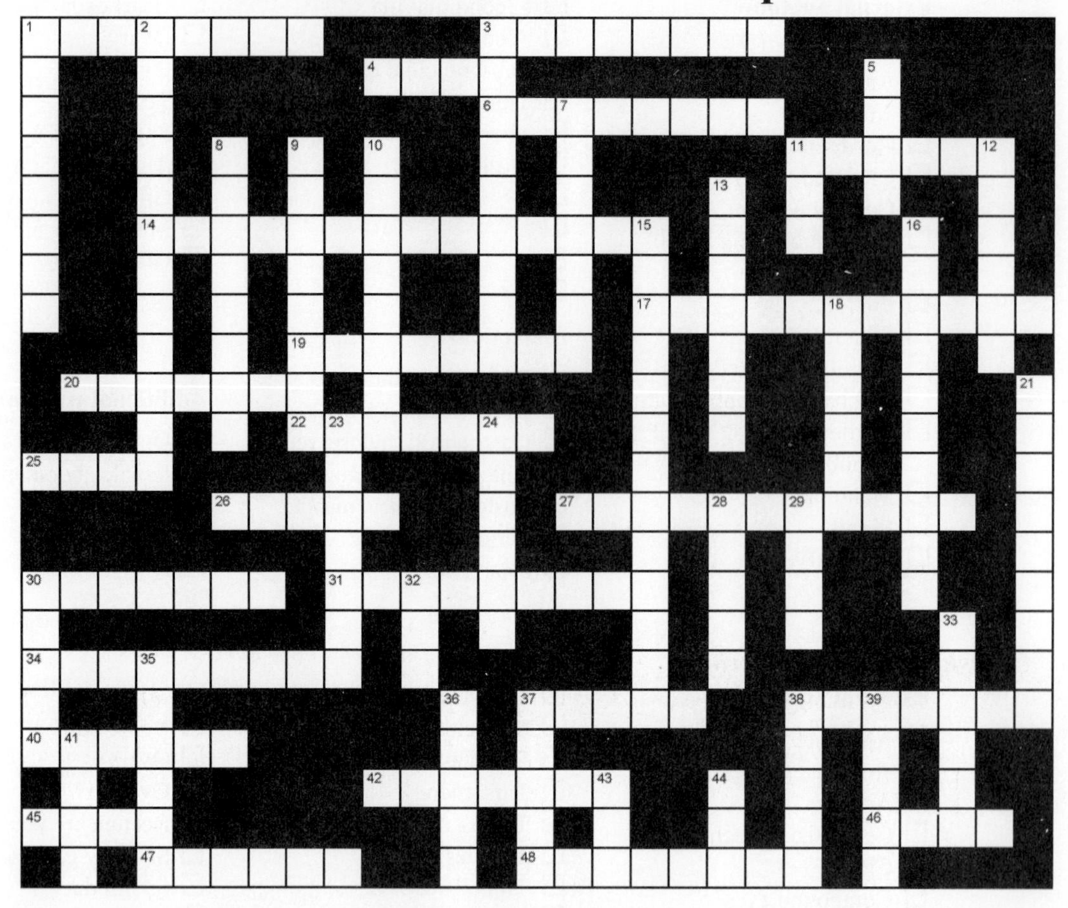

Puzzle solution. An interactive web based version of this puzzle, and its solution, are available on the *Digital Zoology* web site at www.mhhe.com/DigitalZoology/Students. With the interactive puzzle you can check to see if individual words or the whole puzzle is correct, and get hints for single letters.

Across

1 Describes the appendages of crustacea. (8)

3 The feeding appendage found in Uniramia and Crustacea. (8)

4 Because the grinding surface of the Crustacean mandible is located here on the mandible, we refer to it as a gnathobasic mandible. (4)

6 These strengthen and help prevent an insect's trachea from collapsing. (8)

11 Unlike other arthropods, the compound eyes of crustacea are found on these. (6)

14 A synonym for incomplete metamorphosis is. (14)

17 This part of the procuticle is not chemically cross-linked in an arthropod. (11)

19 The main nitrogenous waste produced by insects. (4,4)

20 These large molecules become chemically cross-linked in the exocuticle of an arthropod. (7)

22 The uniramous appearance of the walking legs of a crayfish is referred to as being this. (7)

25 Ancestrally these were found on every segment of an arthropod. (4)

26 Because of the cuticle, in arthropods muscles are arranged this way rather than in sheets. (5)

27 The two arms of a crustacean's biramous appendage are attached to the body by this part of the limb. (11)

30 This is the lower of the two cuticular plates that surrounds the crayfish abdomen. (7)

31 This is inside a statocyst. (9)

34 The only living part of the arthropod exoskeleton. (9)

37 How many teeth are there in the stomach of a crayfish? (5)

38 Statocysts detect this. (7)

40 An insect's upper lip. (6)

42 An insects's Malpighian tubules empty their contents into this part of the gut. (7)

45 You won't find this stage in a hemimetabolous insect. (4)

46 The Arthropoda is considered to be a very successful group in part because of its _____ evolutionary history. (4)

47 This insect mouthpart is unlike the rest because its paired appendages are fused in the middle. (6)

48 Describes the separation of the old cuticle from the epidermis. (8)

Down

1 This nonliving membrane is found at the base of an arthropod's cuticle. (8)

2 Extremely fast wing beat frequencies are possible in insects because of this type of muscle associated with flight. (12)

3 For insects that don't drink water, this type of water is the only source that they have. (9)

5 The main sensory tagma of an insect. (4)

7 The innermost part of a biramous appendage (7)

8 This is the largest body cavity in an arthropod. (8)

9 The unique image forming sensory apparatus of an arthropod is this type of an eye. (8)

10 Blood in the dorsal vessel of an insect is pumped toward this part of the animal. (8)

11 Based on neurological evidence, the uniramian head consists of this many segments. (3)

12 In addition to large eyes, insects also have these eyes. (6)

13 Originally each of the crustacean limbs was involved in gas exchange, locomotion, and this. (7)

15 A male insect's sperm package. (13)

16 Of the four major groups of arthropoda, these are extinct. (10)

18 The most posterior appendage on a crayfish. (6)

21 Glues that attach the eggs to something after they are laid are produced by these glands in the female. (9)

23 A more technical term for moulting, which also describes the escape of an arthropod from its old cuticle. (7)

24 The outermost part of a biramous appendage. (6)

28 The name of the openings in the crustacean hearts. (5)

29 Arthropods have this type of circulatory system. (4)

30 Insects protect the egg from drying out by covering it with a this. (5)

32 A terrestrial insect may swallow this to help escape from the old cuticle. (3)

33 The carbohydrate polymer found in arthropod cuticle. (6)

35 The side of the insect where you would find a tergite. (6)

36 Although almost all insects have them now, ancestrally they didn't. (5)

37 In arthropods, metameres combine to form these larger functional body units. (5)

38 Large crustaceans use these to breathe; it's not a problem for the smaller ones who use simple diffusion. (5)

39 Wings are found only in this stage of the insect life cycle. (5)

41 This opening is found in the telson of a crayfish. (4)

43 Crustacea have this many pairs of sensory appendages on their heads; it's a crustacean trait. (3)

44 Is it true that even with an open circulatory system, crayfish have veins and arteries? (3)

Self Test - Arthropoda

Use the following labels to identify the photographs. You may have to use a label more than once, and some labels may not be appropriate for the photographs. Answers are available on the *Digital Zoology* web site at www.mhhe.com/digitalzoology. Be sure to try the interactive Drag-and-Drop quizzes that are available on the *Digital Zoology* CD-ROM. A color version of this Self Test is available in the Adobe Acrobat version of the Student Workbook in the workbook folder on the *Digital Zoology* CD-ROM.

Specimens

A- Crayfish

B- Grasshopper

Labels

1) Abdomen
2) Abdominal artery
3) Abdominal muscle
4) Antenna
5) Branchial chamber
6) Cephalothorax
7) Cercus
8) Cheliped
9) Collaterial glands
10) Compound eye
11) Crop
12) Digestive ceca
13) Digestive gland
14) Fat body
15) Femur
16) Heart
17) Hindgut
18) Intestine
19) Malpighian tubules
20) Mesothoracic leg
21) Midgut
22) Pronotum
23) Prothoracic leg
24) Tarsus
25) Tibia
26) Uropod
27) Walking leg
28) Wing

ECHINODERMATA

Inside *Digital Zoology*

As you explore the echinoderms on the *Digital Zoology* CD-ROM, don't miss these learning tools:

Photos of dissected specimens of the preserved sea star and sea urchin and a look at brittle stars, sand dollars, and some echinoderm fossils.

Drag-and-drop quizzes on the key structures of the sea urchin and sea star.

Interactive cladogram showing the major events that gave rise to the five living echinoderm classes. A summary of the key characteristics of each class are combined with an interactive glossary of terms.

Defining Differences

Some of the differences described in the following sections are features that arise for the first time in the Echinodermata, and they define the phylum. Others are things that are important for understanding how the animals in the phylum function. Whichever is the case, you'll want to watch for examples of the following in the various echinoderm specimens that you will be examining.

- Pentaradiate symmetry with unique
- tube feet and water vascular system, an
- endoskeleton, and protection by
- pedicellaria.

Pentaradiate symmetry

In a mixed basket of invertebrates you would probably have no trouble picking out most of the echinoderms. Not because you knew their names, but because of their **radial symmetry,** most often based on fives, called **pentaradiate symmetry**. At one time biologists put echinoderms and cnidarians into a single **taxon** — the Radiata. The assumption was that the cnidarians had the **diploblastic** radial plan that ultimately evolved into the **triploblastic** echinoderms. It made sense until investigators found that echinoderm larvae were actually **bilaterally symmetric** and underwent **metamorphosis** to become radially symmetric.

✍ *What are the different types of symmetry that animals can have?*

There are distinct advantages to being a bilaterally symmetric animal, and most animal body plans have this type of symmetry. But, for some reason, the echinoderms chose to do things differently and returned to radial symmetry. There must have been some advantage for doing that, and to find out what it was we have to take a look at the origins of the phylum.

Like a variety of other ancient animals that existed at that time, sessile sponges and cnidarians for example, ancestral echinoderms were **suspension feeders** who trapped and captured food from the water that surrounded them. Echinoderms developed a unique way to catch their food using the tube feet that lined each of the arms that surrounded their central mouth. Once caught, the food was moved to the mouth by a combination of tube feet and a ciliated groove that ran the length of the arm. A new way of suspension feeding was developed and along with it the beginnings of the echinoderms.

✍ *How do sponges and cnidarians trap and capture their food?*

It worked well for them, and when they gave up their sessile existence to move across the substrate, echinoderms kept their basic radial symmetry. As you'll see when you look at echinoderms in your lab, or in *Digital Zoology*, this has been modified in a variety of ways. There are even some echinoderms that show signs of returning to bilateral symmetry!

Tube feet and the water vascular system

Like all deuterostomes, echinoderms have a **tripartate coelom.** Part of the **mesocoel** has been used to form the unique **water vascular system,** which includes the tube feet that extend from the oral side of an echinoderm. Each **tube foot** is connected to the rest of the water vascular system and ultimately to external sea water through the **madreporite** found on the aboral side of the animal. Individually each tube foot can be hydrostatically isolated from the rest of the system allowing it to function as a small **hydrostatic skeleton**. A combination of muscle contractions in the wall of the tube foot and the suction cup action of the tip, allows each of the tube feet to pull the echinoderm along. It might seem inefficient, but with thousands of tube feet working together, echinoderms are able to move, although slowly. Few echinoderms display the ancestral use of the tube feet in feeding. Modern echinoderms have their aboral surface facing down, and the tube feet are used for locomotion rather than feeding.

✍ *What are the different parts of the water vascular system, and how are they connected to each other?*

Endoskeleton

Echinoderms get their name from their spiny skin and their spines, made of **ossicles,** and embedded in the **mesoderm,** creating the **endoskeleton**. What makes an echinoderm's endoskeleton unique is how it's made of calcareous ossicles that start as single calcite spicules. Gradually, as more calcite is deposited, an ossicle takes on its final shape. They aren't solid; instead they're a hollow mesh of crystallized calcite filled with the living tissue. Their different shapes are key characteristics that echinoderm taxonomists use to identify different classes and species in the phylum. Ossicles may be small or fuse to form large skeletal plates.

Pedicellaria

The pincerlike **pedicellaria** found on the surface of some echinoderms are good examples of highly modified ossicles. It was once thought that these were some sort of **ectoparasite**, especially because they have their own musculature and react to their environment without any connection to the rest of the animal. Their role is debatable. Some prevent other animals from sitting on the echinoderm, suggesting a defensive role. There are even pedicellaria that inject toxic poisons to help protect the echinoderm. Still others seem to be responsible for keeping the surface of the animal clean and clear of any debris that settles on the animal. After all, these aren't the fastest animals in the oceans and anything falling through the water column has the potential to land on them.

A Closer Look Inside *Digital Zoology* - Echinodermata

Sea star

The sea star, *Asterias*, is a good introduction to the general features of the phylum Echinodermata. Try rubbing your fingers across the aboral surface of the specimen. You'll feel the tiny spines and understand how the phylum got its name. Even larger ambulacral spines can be found on the oral surface where they protect the delicate tube feet found in the groove. The tube feet and the madreporite on the aboral surface are the only parts of the water vascular system that you can see externally. You can see the rest when you dissect the specimen.

Internally, each arm contains the digestive glands and gonads, surrounded by a spacious perivisceral body cavity. The proximity of the digestive gland to the gonad is handy, allowing ingested nutrients to pass quickly to the gonads, transported by the fluids of the **perivisceral** cavity. Fluid in the perivisceral cavity is an important circulatory fluid that not only carries nutrients to the gonads and other tissues, but it also absorbs oxygen through fingerlike projections of the wall of the cavity, called the **dermal branchia**, which extend through the body wall to the outside.

The dermal branchia aren't the only places where gas exchange occurs. The extensive surface of all of those tube feet is equally important, and oxygen diffuses across them to the fluid that fills the water vascular cavity. Exchange with the perivisceral fluid occurs across the ampullae of the feet, which extend into the main body cavity. The water vascular system and perivisceral body cavity are lined with cilia make sure that the contents of both are always well mixed.

Just as oxygen can diffuse in, so too can nitrogenous wasters diffuse out, and both dermal branchia and tube feet are important surfaces for removing these wastes. As you can see, the importance of tube feet in how the sea star functions goes beyond locomotion.

Inside the central disc, the first of two stomachs, the **cardiac stomach**, can turn inside out when the animals feed. As you might imagine, being a predatory echinoderm is no easy task when most of the animals around you move considerably faster. But there are a few animals that are potential prey, and one of them is bivalve clams. The closed shells of these molluscs are usually effective protection from most predators but the sea star is able to take advantage of one of the minor flaws in a clam's seemingly impervious armor. The shells rarely close perfectly, and there are small gaps, or spaces, along the edges where the two shells meet. The sea star surrounds the clam, holding the edges of the shell against its mouth. The sea star then everts the membranous cardiac stomach squeezing it through these small openings in the clam's closed shell. Digestive enzymes released into the clam start to break down the clam's tissues and cilia lining the sea star's digestive tract transport the nutrients into the second stomach, the **pyloric stomach**, and out to the pyloric ceca in each arm.

Sea urchin and sand dollar

The sea urchin and sand dollar have modified their original body plans but in two different ways — by either forming a hollow globe (regular) or a flattened solid disc (irregular).

Sea urchins

With a sea urchin you have to look closely to see what remains of the five ambulacral regions. The surface of the spherical **test** is alternating ambulacral and interambulacral regions with the long tube feet extending past the spines that cover the animal. Both the tube feet and the spines are important for locomotion. The ball-and-socket joint that anchors each spine to the surface of the test is one of the rare occurrences of this type of joint in invertebrates.

Protruding from the center of the aboral surface, you can see the tips of the five teeth of Aristotle's lantern formed by the same process that makes the endoskeleton. In this case, the ossicles form complex pyramid-shaped plates, each with an interlocking tooth. Musculature attached to the lantern allow the plates to move relative to each other, and the entire lantern can open, close, and rock back and forth as the urchin feeds. Surrounding the mouth are a variety of different types of **pedicellaria,** many of which are stalked.

Most urchins are herbivores, and one of things you'll notice when you open your urchin is the size of the intestine that winds its way through the test. It starts in one direction, then reverses direction as it makes its way to the anus at the top of the aboral surface. Like any other animal that has adopted an herbivorous existence, their gut is extensively modified to allow for processing large quantities of this nutrient-poor food, which explains its length. There is another unique modification that you'll see in the aboral part of the intestine. A **siphon** takes the water from the ingested meal and shunts it to a more distant part of the gut. This way the urchin is able to concentrate the food it consumes. The perivisceral cavity of the sea urchin is large, and suspended from the aboral surface are the five pairs of gonads.

✍ *Why are plants considered a nutrient-poor food?*

Sand dollars

Sand dollars are specialists at burrowing, and the long spines found in the sea urchins are considerably smaller in these animals. The five **petaloids** of the ambulacral regions appear on the aboral surface of the animal, and in sand dollars the tube feet are used primarily for gas exchange rather than locomotion. Sand dollars show signs of the return of bilateral symmetry. The oral opening is often positioned at one end of the disk rather than at the center, making it hard to define oral and aboral surfaces. The openings in the test are **lunnules,** and they also examples of how sand dollars are returning to bilateral symmetry. The function of the lunnules is not certain. They may either stabilize the animals against water flow or may channel the food to the mouth of the animal.

Brittle stars and basket stars

Brittle stars get their name from their delicate arms, which easily break when they are handled. In these animals, ossicles in each of the arms form interlocking vertebrae, which can bend against each other in much the same way that the vertebral column does in vertebrates. Lateral extensions of the ossicles prevent individual vertebrae from slipping past each other, and bands of muscles that run the length of the arm allow it to easily bend and twist. Tube feet are no longer involved in locomotion. Instead, serpentine movements of the flexible arms are how these echinoderms move. It's a modification that makes them the most mobile in the phylum. If you look on the aboral surface of the arms, you won't find any tube feet; you won't even find an ambulacral grove. The central disc in the brittle star is distinct, and if you look at the aboral surface, you'll see the bursal slits, the openings to where the gonads are located.

The ancient history of the phylum can be seen in the fossils of the basket stars and some of their relatives. You can see the characteristics of the phylum, including the branched arms and their **pinnules**, which increased the surface area for capturing food.

Cross-References to Other McGraw-Hill Zoology Titles

Integrated Principles of Zoology, 11th edition. C.P. Hickman, L.S. Roberts & A. Larson. Chapter 23.

Animal Diversity, 2nd edition. C.P. Hickman, L.S. Roberts & A. Larson. Chapter 13.

Zoology, 5th edition. S.A. Miller & J.P. Harley. Chapter 16.

Biology of the Invertebrates, 4th edition. J. Pechenik. Chapter 20.

Laboratory Studies in Integrated Principles of Zoology, 10th edition. C.P. Hickman, F. Hickman & L. Kats. Chapter 16.

General Zoology Laboratory Guide, 13th edition. C. Lytle. Chapter 14.

Structures Checklist

Here are some of the structures that you should be able to easily find in *Digital Zoology* and the specimens that you will be looking at in your lab. After reading your lab handout, you might want to add more and, depending on the equipment available in your lab, you might see more. As you study the material, you might also want to make some notes on how some of these structures looked or include a drawing in your lab notes. (Structures indicated by * may be hard to see.)

Sea star

External

- ☐ Aboral surface
- ☐ Ambulacral groove
- ☐ Ambulacral spines
- ☐ Anus*
- ☐ Arms - bivium
- ☐ Arms - trivium
- ☐ Central disk
- ☐ Dermal branchia
- ☐ Eye spot*
- ☐ Madreporite
- ☐ Mouth
- ☐ Oral surface
- ☐ Pedicellaria
- ☐ Spines
- ☐ Tube feet
- ☐
- ☐
- ☐

Internal

- ☐ Ambulacral ossicles
- ☐ Ampullae of the tube feet
- ☐ Cardiac stomach
- ☐ Genital pore*
- ☐ Gonads
- ☐ Lateral canal
- ☐ Pyloric duct
- ☐ Pyloric (digestive) ceca
- ☐ Pyloric stomach
- ☐ Radial canal
- ☐ Ring canal *
- ☐ Stone canal
- ☐
- ☐
- ☐

Additional structures

- ☐
- ☐
- ☐

Sea urchin

External

- ☐ Aboral surface
- ☐ Ambulacral region
- ☐ Anus
- ☐ Buccal podia
- ☐ Genital plates
- ☐ Interambulacral region
- ☐ Madreporite plate
- ☐ Mouth
- ☐ Oral surface
- ☐ Pedicellaria
- ☐ Peristomial gills
- ☐ Spines
- ☐ Teeth of Aristotle's lantern
- ☐ Tube feet
- ☐
- ☐
- ☐
- ☐

Internal

- ☐ Aboral intestine
- ☐ Ambulacral region
- ☐ Ambulacral plates
- ☐ Ampullae of the tube feet
- ☐ Aristotle's lantern
- ☐ Compass
- ☐ Dental sac
- ☐ Epiphysis
- ☐ Esophagus
- ☐ Gonads
- ☐ Interambulacral region
- ☐ Lantern protractors
- ☐ Lantern retractor
- ☐ Oral intestine (stomach)
- ☐ Pyramid
- ☐ Rectum
- ☐ Ring canal*
- ☐ Radial canal*
- ☐ Siphon
- ☐ Tooth
- ☐
- ☐
- ☐
- ☐
- ☐
- ☐

Additional structures

- ☐
- ☐
- ☐

Sand dollar

External

- ☐ Ambulacral region
- ☐ Food groove
- ☐ Genital plates
- ☐ Genital pores
- ☐ Interambulacral region
- ☐ Lunnules
- ☐ Madreporite
- ☐ Oral surface
- ☐ Position of the anus
- ☐ Position of the mouth
- ☐ Spines
- ☐
- ☐
- ☐
- ☐
- ☐
- ☐

Additional structures

- ☐
- ☐
- ☐
- ☐
- ☐

Brittle star

External

- ☐ Oral surface
- ☐ Aboral surface
- ☐ Central disk
- ☐ Arm spines
- ☐ Radial shield
- ☐ Bursal slits
- ☐ Mouth
- ☐ Jaws
- ☐ Buccal tube feet
- ☐
- ☐
- ☐
- ☐

Additional structures

- ☐
- ☐
- ☐
- ☐
- ☐

Crossword Puzzle — Echinodermata

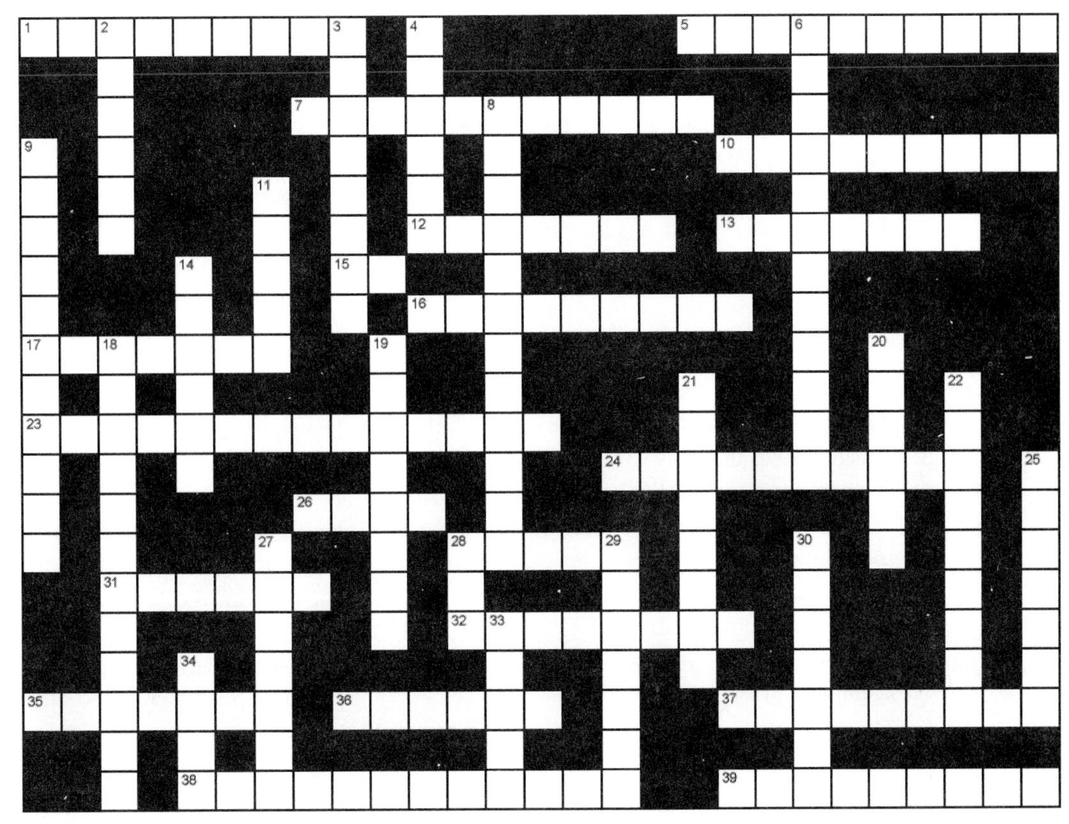

Puzzle solution. An interactive web based version of this puzzle, and its solution, are available on the *Digital Zoology* web site at www.mhhe.com/DigitalZoology/Students. With the interactive puzzle you can check to see if individual words or the whole puzzle is correct, and get hints for single letters.

Across

1 The pincerlike structures on the surface of the sea star were once thought to be these, not a part of the sea star. (9)

5 Like all deuterostomes, echinoderms have this type of divided coelom. (10)

7 Water enters the water vascular system through openings in this structure. (11)

10 Whose lantern was used to name the unique sea urchin feeding apparatus? (9)

12 The radial canals of an echinoderm's water vascular system are connected to these canals, and then to the tube feet. (7)

13 The ancestral echinoderms tube feet were originally used for this. (7)

15 Ancestrally, the mouth of an echinoderm faced in this direction relative to the substrate to which it was attached. (2)

16 In brittle stars the ossicles in the arms resemble these. (9)

17 This part of the tube foot extends into the body cavity of an echinoderm. (7)

23 In sea stars, gas exchange occurs in the tube feet and these. (6,8)

24 The tube feet in an echinoderm extend from this groove. (10)

26 In echinoderms, the blastopore ultimately forms this structure. (4)

28 Sea urchin shells are also called this. (5)

31 In urchins, a ball-and-socket joint connects these to its shell. (6)

32 The calcareous plates that form the internal skeleton of an echinoderm. (8)

35 The stomach closest to the mouth in a starfish. (7)

36 Urchins concentrate their ingested meal by removing the water from it using this modification of the digestive tract. (6)

37 The flat echinoids, which includes urchins and sand dollars, are referred to as being this. (9)

38 An echinoderm's skeleton is this type. (12)

39 Absence of this system has limited echinoderms to the marine environment. (9)

Down

2 In an echinoderm, the ring canal connects to this canal before connecting to the tube feet. (6)

3 This describes the body cavity of a sea urchin compared to a sand dollar. (8)

4 In brittle stars the gonads are inside these special pouches. (6)

6 These defensive structures cover the surface of some echinoderms. (12)

8 Echinoderms have this special type of radial symmetry. (12)

9 The arms of a sea star are connected to this part of the animal. (7,4)

11 These keep the fluids on both echinoderm body cavities moving. (5)

14 When present these vesicles act as reservoirs that store water for the water vascular system. (6)

18 The name of the body cavity that surrounds the internal organs in an echinoderm. (12)

19 These holes let water pass from the upper to the lower surface of a sand dollar. (8)

20 Without bilateral symmetry, we refer to the main surface opposite the mouth in echinoderms as this side of the animal. (6)

21 The main locomotory structures used by echinoderms. (8)

22 A larval echinoderm has this type of symmetry. (9)

25 The globe-shaped echinoids, which includes urchins and sand dollars, are referred to as being this. (7)

27 This term the circulatory and nervous systems in echinoderms. (7)

28 The number of digestive ceca in each arm of a starfish. (3)

29 In an echinoderm the tip of the tube foot uses this to help it attach to the substrate. (7)

30 This stomach connects to the digestive glands found in each of a sea star's arms. (7)

33 This canal connects the madreporite to the ring canal in an echinoderm. (5)

34 Most echinoderms have this number of arms. (4)

Self Test - Echinodermata

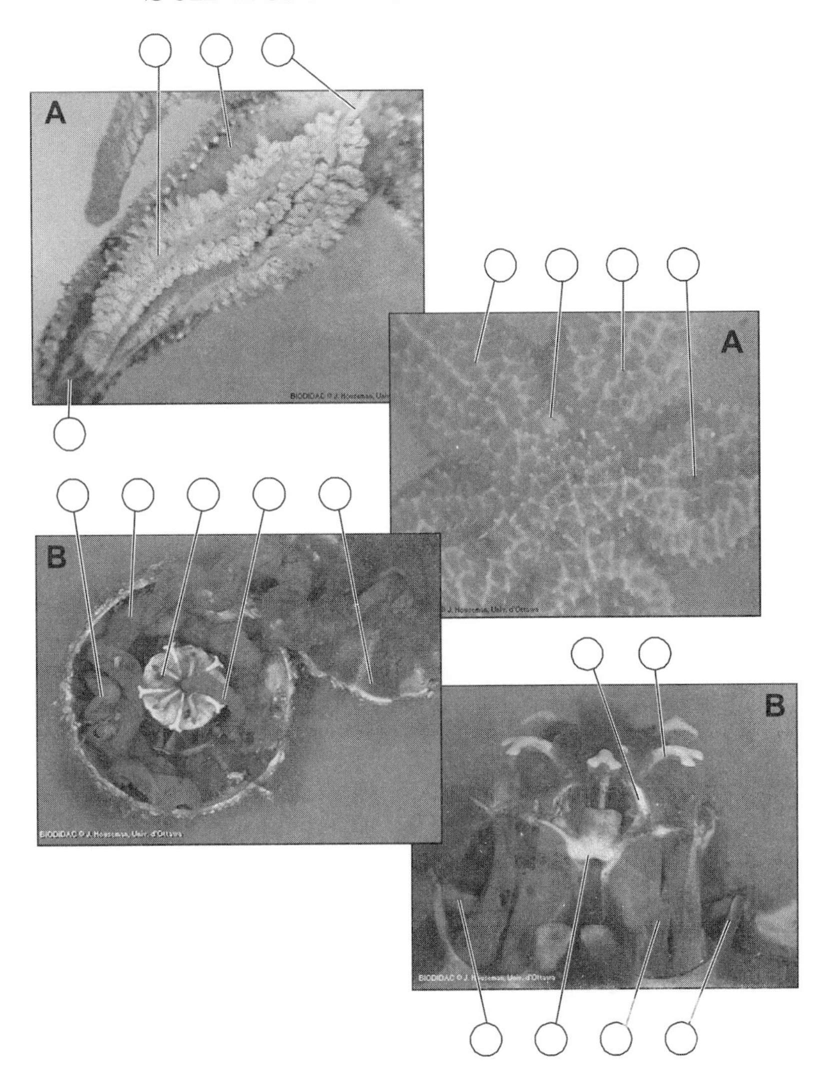

Use the following labels to identify the photographs. You may have to use a label more than once, and some labels may not be appropriate for the photographs. Answers are available on the *Digital Zoology* web site at www.mhhe.com/digitalzoology. Be sure to try the interactive Drag-and-Drop quizzes that are available on the *Digital Zoology* CD-ROM. A color version of this Self Test is available in the Adobe Acrobat version of the Student Workbook in the workbook folder on the *Digital Zoology* CD-ROM.

Specimens

A- Sea star
B- Sea urchin

Labels

1) Aboral intestine
2) Ambulacral ridge
3) Aristotle's lantern
4) Auricle
5) Bivial arm
6) Compass
7) Dental sac
8) Epiphysis
9) Esophagus
10) Gonad
11) Lantern protractors
12) Lantern retractors
13) Madreporite
14) Oral intestine
15) Pyloric cecum
16) Pyloric duct
17) Trivial arm

CHORDATE ORIGINS

Inside Digital Zoology

Explore the origins of the Phylum Chordata on the *Digital Zoology* CD-ROM, using these learning tools:

- Photos of preserved specimens and cross-sections of the hemichordate *Balanoglossus*, whole mounts and cross-sections of larval and adult urochordates, a cephalochordate, the lancelet *Branchiostoma*, and the ammocoete larva of the lamprey all help to understand the origins of the chordates.

- Interactive cladogram showing the major events that gave rise to the main deuterostome groups. A summary of the key characteristics of each of the deuterostome phyla and subphyla of chordates and hemichordates are combined with an interactive glossary of terms.

Defining Differences

To understand the origins of the phylum Chordata we look at the Hemichordata, which preceeded them; the protochordates, which include the Urochordata and Cephalochordata; and the ammocoete larval stage of the lamprey. When examining the specimens in the lab or in *Digital Zoology*, these are features that define the phylum and are important for understanding how animals in the phylum function. You'll want to watch for examples of these in the various specimens that you will be looking at in your lab or in *Digital Zoology*.

- A notochord,
- postanal segmentation,
- dorsal hollow nerve cord, and
- pharyngeal gill slits and the process of
- paedomorphosis explaining their origins.

Notochord

The **notochord** gives animals in the phylum Chordata their name and is another example of an innovation in body plans that gave rise to a new group of animals. When you look at a notochord in your lab or in *Digital Zoology*, it seems to be such a simple structure to have had such an important role in the origins in the phylum – a fluid-filled rod of tightly packed cells surrounded by a sheath of elastic connective tissue. You might be more surprised to find that the character that defines the phylum is only present in the most primitive subphyla: urochordates and cephalochordes. In vertebrates the notochord's role as a skeletal structure has been replaced by the vertebral column.

To understand the significance of the notochord you also need to know about the muscles associated with it and how the combination of the two resulted in this new way of moving. Chordates' muscles are arranged segmentally, as **myomeres,** along the length of the notochord with each myomere connected to its neighbor by shared **myosepta**. Muscle fibers in the myomere are aligned in only one direction, anterior to posterior, and there's no fluid-filled coelomic space to act as a **hydrostatic skeleton**. The result is that in the absence of a notochord, when muscle in a myomere contract, it shortens the length of the myomere and stretches the adjacent myomere connected to it by the myosepta. This creates accordionlike ripples down the side of the animal.

When we add the notochord the contraction of the myomeres bends the notochord. When the muscles relax, the elasticity of the notochord returns it to its original straight position, stretching the myomere back to its original length. Repeat the process with the myomeres on the opposite side of the animal and the notochord bends in the opposite direction. When the myomeres relax, the notochord straightens again. By coordinating the contraction of the myomeres down the length of the notocord, the result is the swimming motion typical of the aquatic animals in the phylum with their trunk and tail undulating from side to side, propelling the animal forward. This is a new way of moving.

Postanal segmentation

There's a little more to this than a combination of a notochord and myomeres. The myomeres and notocord extend beyond the anal opening and add more length to this now propulsive structure. This additional length in the body is referred to as **postanal segmentation,** which may sound like a rather technical term for what we're talking about, the first animals with tails. In the other animals that we've seen, the anal opening was always found on the last segment, or most posterior part of the body; this is not so in chordates.

Dorsal hollow nerve cord

Another of the unique features of the chordates is their dorsal hollow nerve cord, which forms from the two sides of the neural plate that fuse together to become the nerve cord lying on top of the notochord and innervating the myomeres that run along its length.

Pharyngeal gill slits

Chordates share their ancestral feeding habit, suspension feeding, with many of the first animals that appeared around the same time. So how do they do it differently? As you might expect, cilia are involved, and they propel the particle-laden water into the digestive tract. However, it's in the most anterior part of the tract, the **pharynx**, where the difference occurs.

In chordates the pharynx is perforated with openings so that water can pass from inside the pharynx through the pharyngeal openings, the slits, to the outside of the animal. The cilia on the walls of the pharynx are moving the water and at the same time trapping food particles contained in the water. Particulate food isn't only trapped by the cilia on the pharynx. An **endostyle** on the ventral surface of the pharynx will also catch food in its mucous net.

Paedomorphosis

What are the events that led to the first chordates, and what evidence is there to explain how it happened? The proposal made by Garstang in 1928 is still generally accepted and identifies the tunicates as being important in explaining the origins of the group. Tunicates have a **dimorphic life cycle** with a sessile reproductive adult and an immature larval tadpolelike stage. One of the important steps in the evolution of the chordates, as proposed by Garstang, is what he called **paedomorphosis**, which involves the immature larval form becoming the reproductive stage and the ancestor to the phylum.

✍ *What are the other stages, or steps, in Garstang's theory on the origins of the Chordata?*

A Closer Look Inside *Digital Zoology* - Chordate Origins

Balanoglossus: *The acorn worm*

There are two hemichordate classes. There are burrowing animals, such as the acorn worm, Enteropneusta, and the sessile colonial forms, such as the Pterobranchia. Not everyone agrees that the hemichordates should be a phylum, and there is increasing evidence that they are **polyphyletic** and that the two classes should be elevated to phyla. In this scenario the Enteropneusta would be more closely related to the chordates, the Pterobranchia more closely related to echinoderms.

Like all deuterostomes the acorn worm has a **tripartate body,** and the proboscis, collar, and trunk have cavities inside of them that correspond to the three coelomic spaces. You'll be able to see the coelomic spaces in the cross-section slides. The proboscis of an acorn worm is important for both locomotion and feeding, and the musculature inside it and the coelomic space work like a **hydrostatic skeleton** as the worm pushes into the soft substrate.

Acorn worms are **mucociliary feeders**, and particulate food collected directly from the water, or the soft substrate that the proboscis has been pushed into, is trapped in mucous, which is then propelled to the mouth by cilia. The mouth is located at the edge of the collar, and when it's open, food-laden mucous passes into the pharynx. The wall of the pharynx has U-shaped **gill slits** that open to the outside through **gill pores.** The walls of the slits, **gill bars**, are lined with cilia, which keep the food inside the pharynx while removing the water, concentrating the food before it enters the digestive tract. The pharyngeal gill slits are similar to those of the protochordates, although in enteropneusts they may not have a respiratory role, being primarily involved in feeding.

Pharyngeal gill slits aren't the only things that acorn worms seem to share with the protochordates. Their **stomochord** may be similar to a **notochord**. Embryologically the notochord forms as an evagination along the length of the embryonic gut. In the acorn worm the stomochord arises in a similar way, but only at the most anterior end of the gut, suggesting that there may some similarity between the two. Acorn worms also have a hollow neurochord in the collar that some speculate may be related to the hollow dorsal nerve cord of the chordates.

Urochordata: The tunicates

As mentioned tunicates have a **dimorphic life cycle** with the reproductive adult being sessile and the larval stage being free swimming. The urochordates get their common name, tunicates, from the covering, **tunic**, of the sessile adult and the name of the phylum from the notocord found only in the tail of the urochordate larva. Other than the tunic, the other obvious features of the adult are the incurrent and excurrent siphons. Tunicates are suspension feeders, and inside the incurrent siphon is an enlarged perforated **pharynx** that forms a **branchial basket**. This is a ciliated structure, and the **cilia** on the surface of the basket pull the water into the pharynx and across the **pharyngeal slits** where it is trapped in mucous produced by the **endostyle**. Like the acorn worm these are **mucociliary** feeders.

To see the remainder of the vertebrate characteristics we need to take a look at the larval stage of the tunicate. Think of it as a smaller version of the branchial basket that we saw in the adult, but this time it's attached to that new way of swimming that appeared in chordates. The tail of the urochordate larva has the **notocord**, a **dorsal hollow nerve cord,** and small blocks of myomeric muscles that create the undulating swimming motion of the chordates. All of this is located behind the opening of the excurrent siphon of the small branchial basket where the anus is located. It qualifies as **postanal segmentation**, the chordate tail.

Branchiostoma*: The lancelet or amphioxus*

The lancelet, *Branchiostoma*, or amphioxus as it is more often called, is a good specimen for seeing many of the chordate characteristics. Whole-mount slides and cross-sections are the best way to examine this small fishlike animal. If both aren't available in your lab, they are in *Digital Zoology*.

Segmentation of the body is best seen in the arrangement of the **myomeres**. In amphioxus each of the myomeres is V-, or chevron-, shaped, and overlaps with the ones on either side. The tail isn't very big in these animals but there are segments after the anal opening, and that's all we need to see to identify this chordate character. As you're looking at the tail be sure that you don't confuse the position of the anal opening with the **atriopore** where water that has crossed the pharyngeal openings leaves the animal.

Amphioxus is also a **mucociliary feeder**. **Cilia** on the **gill bars** between the gill slits and on the **wheel organ** help to pump water, and the particulate food in it into the mouth. Once the water current enters the **pharynx,** mucous secreted by the **endostyle** traps the particulate materials, and the food-laden mucous is propelled by pharyngeal cilia up and toward the roof of the mouth and from there to the digestive tract. Once the particulate food has been removed, the water passes between the gill bars; through the **pharyngeal gill slits**; into a cavity, the atrium; and out the atriopore. You'll be able to see some of these details in the whole mounts but cross-sections are better. The endostyle does more than create a mucous net; its cells are capable of binding iodine, and it is often thought to be the precursor of the thyroid gland.

The **notocord** and **hollow nerve cord**, two other chordate traits, are easy to see in the cross-sections and although the notocord is visible in the whole mounts, the nerve cord is a little harder to see. The best way to locate the nerve cord is to use the black photoreceptor cells as landmarks. They are inside the nerve cord. Cephalochordates get their name from the extension of the notocord into the head region.

Blood flow in amphioxus follows the general vertebrate pattern with blood flowing toward the anterior end of the animal inside the ventral aorta. Blood flows through the gill arches to a dorsal aorta and back through the body. As it flows through the gill arches there are multiple blood vessels in each arch, rather than the single blood vessel found in most vertebrates. Unlike vertebrates, there is no heart; instead, the major blood vessels are contractile and used to propel the blood through the system.

Ammocoete larva

The larval stage of the lamprey is called an ammocoete larva, and although it looks somewhat like amphioxus don't forget that it is a vertebrate. Look for the chordate characters in the whole-mount slide and cross-sections. The **hollow nerve cord** is located above the **notochord,** and it's enlarged at the anterior end to form a **brain** with three lobes, each with separate functions. The **forebrain** is olfactory and connects with the nostril, and the **midbrain** is visual and is connected to a pair of eyes, which in the larva aren't functional. These larvae spend most of their time with their heads buried in the mud and instead of using their eyes that have special photoreceptors in the tail to detect light. The **hindbrain** is connected to small **ear vesicles** and the remainder of the nerve cord. The **myomeres** in this animal are much smaller than those of amphioxus and don't wrap around the sides of the body in the same way. You may find them easier to see in the sections rather than the small whole mount.

Ammocoetes also have **pharyngeal gill slits** but their shape and number have changed from what we saw in amphioxus. They're no longer U-shaped, and there are only seven pairs. There is one other fundamental difference. Cilia are no longer involved in pumping water through the pharynx. Instead, pharyngeal musculature and the **velum** change the shape of the buccal cavity to suck in food. Ammocoetes are still **mucociliary feeders**, and as water is forced through the gill slits fóod is trapped in mucous produced by goblet cells on the gill bars. The **endostyle** still secretes mucous that helps to trap food but it is now enclosed in a tube that runs along the floor of the pharynx, and the mucous it produces contains iodinated compounds. During metamorphosis it becomes the thyroid gland of the adult lamprey. Cilia are still important in feeding, and they move food into the digestive tract as they did in the early chordates.

The gill bars are much more elaborate in ammocetes. This is related to their increased importance in gas exchange. Ammocetes have gills with lamellae that increase the surface area available for gas exchange. They also have a two-chambered, ventral heart that pumps blood from the ventral aorta over the gills to the dorsal aorta and out to the body

Cross-References to Other McGraw-Hill Zoology Titles

Integrated Principles of Zoology, 11th edition. C.P. Hickman, L.S. Roberts & A. Larson. Chapters 24, 25 & part of 26.

Animal Diversity, 2nd edition. C.P. Hickman, L.S. Roberts & A. Larson. Chapters 13, 14 & part of 15.

Zoology, 5th edition. S.A. Miller & J.P. Harley. Chapter 17.

Biology of the Invertebrates, 4th edition. J. Pechenik. Chapter 21 & 22.

Laboratory Studies in Integrated Principles of Zoology, 10th edition. C.P. Hickman, F. Hickman & L. Kats. Chapter 17 & part of 18.

General Zoology Laboratory Guide, 13th edition. C. Lytle. Chapter 15.

Structures Checklist

Here are some of the structures that you should be able to easily find in *Digital Zoology* and the specimens that you will be looking at in your lab. After reading your lab handout, you might want to add more and, depending on the equipment available in your lab, you might see more. As you study the material, you might also want to make some notes on how some of these structures looked or include a drawing in your lab notes. (Structures indicated by * may be hard to see.)

The acorn worm

External anatomy

- ☐ Abdominal region of trunk
- ☐ Anus
- ☐ Branchial region of trunk
- ☐ Collar
- ☐ Genital region of trunk
- ☐ Gill pores
- ☐ Longitudinal ridges
- ☐ Mouth*
- ☐ Proboscis
- ☐ Proboscis stalk*
- ☐ Trunk
- ☐
- ☐

Cross section

- ☐ Circular proboscis musculature
- ☐ Collar
- ☐ Collar coelom
- ☐ Dorsal nerve cord
- ☐ Dorsal vessel
- ☐ Glomerulus
- ☐ Heart
- ☐ Longitudinal proboscis musculature
- ☐ Mouth
- ☐ Pharynx
- ☐ Proboscis coelom
- ☐ Stomochord (Buccal diverticulum)
- ☐ Ventral vessel
- ☐
- ☐
- ☐

Additional structures

- ☐
- ☐
- ☐
- ☐
- ☐

Tunicates

Whole mount

- ☐ Adhesive papilla
- ☐ Dorsal fin
- ☐ Endostyle
- ☐ Eye spot
- ☐ Head
- ☐ Nerve cord
- ☐ Notochord
- ☐ Oral siphon
- ☐ Pharynx
- ☐ Statocyst
- ☐ Tail

Ventral fin

- ☐
- ☐

Adult

- ☐ Anus
- ☐ Endostyle
- ☐ Excurrent siphon (atriopore)
- ☐ Incurrent siphon (mouth)
- ☐ Intestine
- ☐ Pharynx
- ☐ Stomach
- ☐ Tunic
- ☐
- ☐
- ☐
- ☐
- ☐
- ☐

Additional structures

- ☐
- ☐
- ☐
- ☐
- ☐

Amphioxis

Whole mount

- ☐ Anus
- ☐ Atriopore
- ☐ Caudal fin
- ☐ Dorsal fin
- ☐ Fin rays
- ☐ Gill bars
- ☐ Gill slits
- ☐ Gonad
- ☐ Hepatic caecum (midgut caecum)
- ☐ Intestine
- ☐ Mouth
- ☐ Myomeres
- ☐ Nerve chord
- ☐ Notochord
- ☐ Oral cirri
- ☐ Oral hood
- ☐ Pigment spots
- ☐ Rostrum
- ☐ Tail
- ☐ Velum
- ☐ Ventral fin
- ☐ Vestibule
- ☐
- ☐
- ☐
- ☐

Cross section

- ☐ Atrium
- ☐ Dorsal aorta
- ☐ Dorsal fin
- ☐ Endostyle
- ☐ Epibranchial groove
- ☐ Fin ray
- ☐ Gill bar
- ☐ Gill slit
- ☐ Gonad
- ☐ Intestine
- ☐ Metapleural folds
- ☐ Myomeres
- ☐ Myosepta
- ☐ Nerve cord
- ☐ Notochord
- ☐ Pharynx
- ☐ Ventral aorta
- ☐ Ventral fin
- ☐
- ☐
- ☐
- ☐
- ☐
- ☐
- ☐

Additional structures

- ☐
- ☐
- ☐
- ☐
- ☐

Ammocoete larva

Whole mount

- ☐ Anus
- ☐ Caudal fin
- ☐ Dorsal fin
- ☐ Endostyle (Subpharyngeal gland)
- ☐ Esophagus
- ☐ Forebrain
- ☐ Gill bars
- ☐ Gill pouch
- ☐ Gill slits
- ☐ Hindbrain
- ☐ Intestine
- ☐ Midbrain
- ☐ Nerve cord
- ☐ Notochord
- ☐ Oral hood
- ☐ Oral papillae
- ☐ Pronephric ducts
- ☐ Tail
- ☐
- ☐
- ☐
- ☐

Cross section

- ☐ Dorsal aorta
- ☐ Ear vesicle
- ☐ Endostyle (Subpharyngeal gland)
- ☐ Gill bars
- ☐ Gill lamellae
- ☐ Gill slit
- ☐ Hyperpharyngeal ridge
- ☐ Hypopharyngeal ridge
- ☐ Kidney
- ☐ Intestine
- ☐ Myomeres
- ☐ Myosepta
- ☐ Nerve cord
- ☐ Notochord
- ☐ Pharynx
- ☐ Typhlosole
- ☐
- ☐
- ☐
- ☐
- ☐
- ☐

Additional structures

- ☐
- ☐
- ☐
- ☐
- ☐

Crossword Puzzle – Chordate Origins

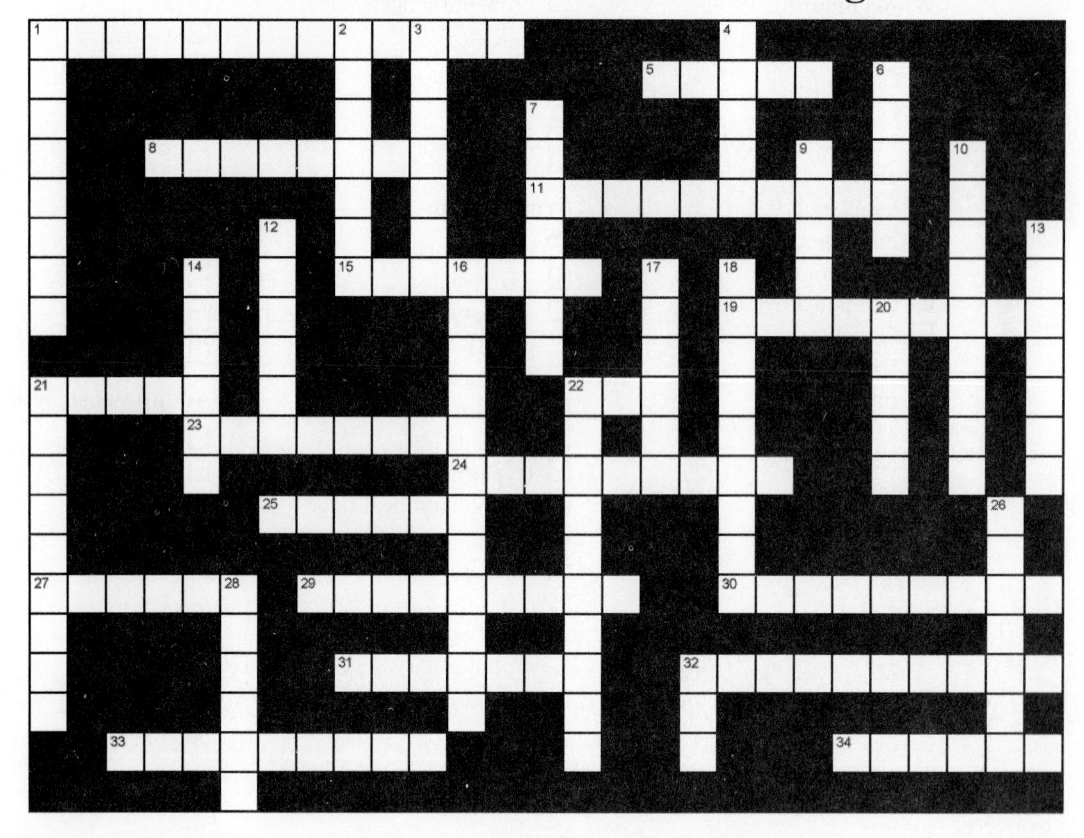

Puzzle solution. An interactive web based version of this puzzle, and its solution, are available on the *Digital Zoology* web site at www.mhhe.com/DigitalZoology/Students. With the interactive puzzle you can check to see if individual words or the whole puzzle is correct, and get hints for single letters.

Across

1 The transformation of the larval into the adult during the life cycle of a urochordate is another example of something we've seen in other animals. (13)

5 Enteropneust hemichordates resemble this type of animal. (5)

8 These support the gill slits in a cephalochordate. (8)

11 These slits help to separate food from ingested water prior to passing the food into the digestive system. (10)

15 Used to describe the V-shaped muscle blocks in a lancelet. (7)

19 In hemichordates the stomocord was mistakenly thought to be this structure and the reason that they were originally included as a subphylum of the chordata. (9)

21 This is the outer covering of an adult urochordate. (5)

22 This structure is U-shaped in a urochordate. (3)

23 The direction in which blood flows in the dorsal blood vessel of an acorn worm (Hemichordata). (8)

24 The name of the lampray larval stage. (9)

25 Position of the heart in an acorn worm (Hemichordata). (6)

27 Water moves through the gill slits into this space before leaving the lancelet. (6)

29 Water leaves the atrium of a cephalochordate through this. (9)

30 The anus of a urochordate is located next to this siphon. (9)

31 Blood in this vessel moves toward the front of a lancelet. (7)

32 Like all deuterostomes, hemichordates have this type of coelom. (10)

33 On the inside of an acorn worm's pharynx these are U-shaped on the inside and connected to pores on the outside. (9)

34 You'll find more of the ancestral chordate characters in this stage of a urochordate's life cycle. (6)

Down

1 The collar of a hemichordate acorn worm is formed from this coelomic space. (8)

2 Phagocytic digestion of ingested food occurs in this cephalochordate cecum. (7)

3 An adult tunicate has this type of existence. (7)

4 Another term that describes that anterior swelling of the dorsal nerve cord. (5)

6 Even though they have gill slits and gill pores, acorn worms don't have these. (5)

7 Urochordates have these; one is incurrent, the other excurrent. (7)

9 Lancelets don't have one of these in their circulatory system. (5)

10 In acorn worms (Hemichordata) the protocoel forms this structure. (9)

12 Common name for a cephalochordate. (6)

13 The larval stage of of a tunicate looks like one of these. (7)

14 In a hemichordate acorn worm this diverticulum of the gut was originally thought to be a notochord. (6)

16 A lancelet pumps its blood through the circulatory system using this structure. (7,5)

17 Branchial arteries are also referred to as this type of arch. (6)

18 This structure on the ventral surface of the urochordate pharynx uses cilia to propel food into the digestive tract. (9)

20 Food trapped on the surface of an acorn worm's proboscis is moved into the gut by these. (5)

21 Urochordates have this common name. (9)

22 The presumed excretory structure of a hemichordate corn worm. (10)

26 The name given to the jawless fish. (7)

28 The nonliving covering that surrounds an adult urochordate is secreted by this. (6)

32 The number of chambers in an agnathans heart. (3)

Self Test – Chordate Origins

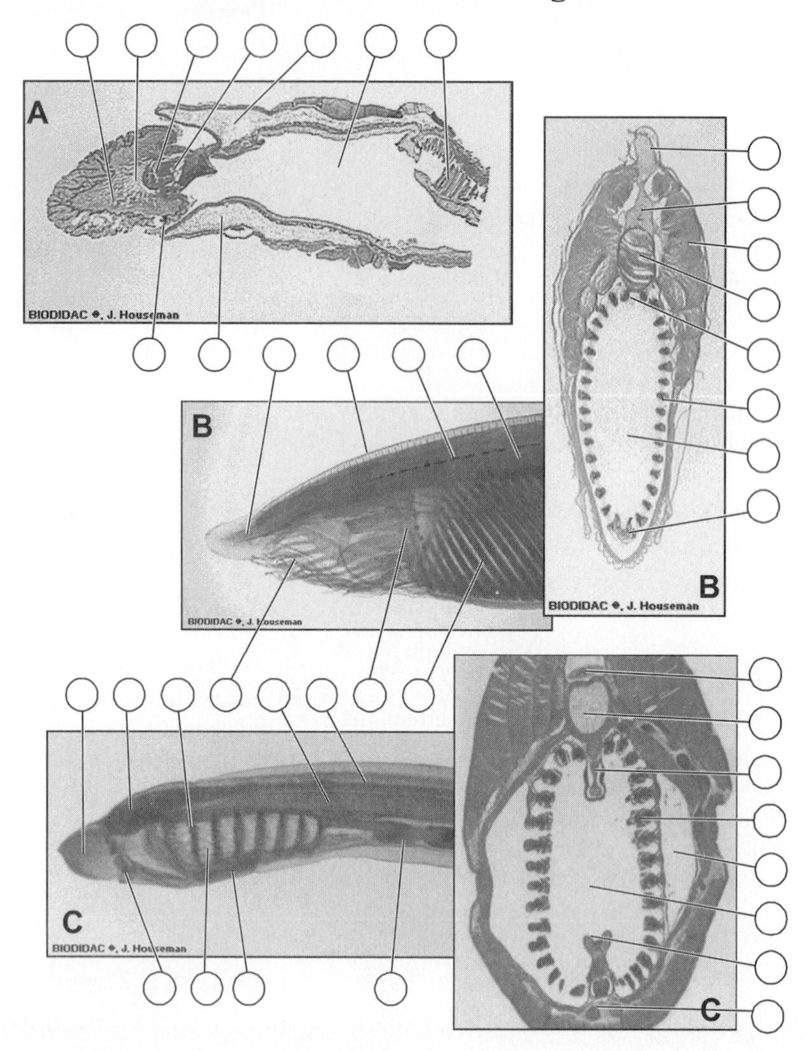

Use the following labels to identify the photographs. You may have to use a label more than once, and some labels may not be appropriate for the photographs. Answers are available on the *Digital Zoology* web site at www.mhhe.com/digitalzoology. Be sure to try the interactive Drag-and-Drop quizzes that are available on the *Digital Zoology* CD-ROM. A color version of this Self Test is available in the Adobe Acrobat version of the Student Workbook in the workbook folder on the *Digital Zoology* CD-ROM.

Specimens

A- *Balanoglossus*
B- *Branchiostoma* - Amphioxus
C- Ammocoete larva

Labels

1) Brain
2) Buccal cirri
3) Collar coelom
4) Dorsal fin ray
5) Endostyle
6) Epibranchial groove
7) Gill
8) Gill bar
9) Gill chamber
10) Gill pouch
11) Gill slit
12) Nerve cord
13) Glomerulus
14) Heart
15) Hyperpharyngeal ridge
16) Hypopharyngeal ridge
17) Mouth
18) Myomere
19) Nerve cord
20) Notochord
21) Oral hood
22) Oral papilla
23) Proboscis muscles
24) Peripharyngeal band (Velum)
25) Pharynx
26) Proboscis coelom
27) Stomocord

JAWED FISHES

Inside *Digital Zoology*

As you explore the jawed fishes on the *Digital Zoology* CD-ROM, don't miss these learning tools:

- Photos of dissected specimens of cartilaginous and bony fishes, which includes the shark and perch, respectively.
- Drag-and-drop quizzes on the external anatomy and skeleton, muscular systems, and internal anatomy for the shark and perch.
- Interactive cladogram showing the major events that gave rise to vertebrate classes. A summary of the key characteristics of each class are combined with an interactive glossary of terms.

Defining Differences

Fish is a pretty broad term that includes anything that's vertebrate, cold-blooded, swims, and has fins. In this part of *Digital Zoology* we take a look at two different types of fish: cartilaginous fish, Chondrichthyes, and bony fish, Osteichthyes. Each of the features described in the following list applies to the two classes and, depending how the feature is modified, it may define differences between the two classes of jawed fishes. Whichever is the case, you'll want to watch for examples of these in the specimens you will be examining.

- Jaws,
- gills,
- paired and unpaired fins with a fusiform body covered in
- scales.

Jaws

We see **jaws** for the first time in living vertebrates with the cartilaginous fish, Chondrichthyes. These, and all other "jawed-mouthed" vertebrates, are collectively referred to as the **gnathostomes**. Their counterparts are vertebrates with a mouth, but no jaws – the **agnathans,** hagfish, and lampreys. Both groups descended from early fishes that were nothing like what we see today. **Ostracoderms,** the agnathan relative, and **placoderms,** ancestor to modern day jawed fishes, were heavily armored with large **dermal skeletal plates** covering their bodies. Although they dominated the oceans at one time, these ancient fish started to disappear as earth became much warmer and drier at the end of the Devonian. (In the next chapter we'll see how these same conditions started the vertebrate transition to land and the origins of the Amphibia.) Whatever the reason is for the disappearance of the armored fish they were replaced by cartilaginous and bony fish that resemble those we see today. As you look at each of the two groups during your lab, or in *Digital Zoology*, you'll see hints about their evolutionary past. Remember the two groups diverged long ago; each is highly specialized. Don't look at the modern day Chondrichthyes as an ancestor to the bony fishes, it didn't happen that way.

Jaws were a major innovation for the first animals that had them. They could feed more aggressively by sucking up large amounts of substrate, trapping and holding prey in place, or taking a bite out of their prey. As we've seen, any innovation that allows for a new way of feeding usually results in diversification of the group. That's what happens with the gnathostomes. So how do biologists believe that jaws first appeared?

As we've seen in the previous chapter, early chordates fed on particulate food pulled into the mouth by **cilia** inside the pharynx. The same cilia filtered the particulate food from the water sending food into the digestive system and water out through **gill slits** in the side of the **pharynx**. The openings in the wall of the pharynx were held open by **pharyngeal arches**.

There's another way to move particle-laden water through the pharynx that doesn't require cilia. The pharyngeal cavity becomes the pump, shrinking to force water out through the gill slits on the sides, then increasing in size to pull water in through the mouth. The elasticity of the pharyngeal arches plays a key role in the mechanism. When the pharyngeal muscles contract, the arches bend, and when the muscles relax, the arches spring back to their original shape. This returns the pharyngeal cavity to its original size and pulls in water. Many of the first fishes, ostracoderms for example, used this pharyngeal pumping strategy to feed. They placed their mouths against the substrate and sucked it in and became substrate, rather than suspension, feeders.

There's still some room for improvement. If the diameter of the mouth remains the same as the pharyngeal cavity pumps, as much water is likely to go out of the mouth as come in the mouth. If the diameter of the mouth is decreased at the same time the pharyngeal cavity decreases in size, then more water would be forced out through the pharyngeal slits than through the mouth. The most anterior pharyngeal arch, already capable of bending, bent more than the rest of the arches and the mouth started to close. We have the beginnings of the rudimentary gnathostome jaw as a modification of the first pharyngeal (gill) arch.

Gills

Pharyngeal arches, like all tissues, are supplied with blood vessels. With oxygenated water flowing past them, the arches became an ideal place to extract oxygen from water. The result was gills with a supporting **gill arch**, a **gill raker** on the inside of the arch, and a pair of **gill filaments** on the outside. Once cilia were no longer involved in feeding, the gill rakers kept the ingested food inside the pharynx and let the water flow out and over the gill filaments. Gill filaments are made of minute **lamella** and are the site of gas exchange. Water flows across these lamellae in the opposite direction to the blood inside, in a pattern referred to as **counter current exchange**.

✍ *Why is countercurrent flow of blood and water better than the concurrent flow of the two fluids?*

Other than swimming with an open mouth and changing the shape of the buccal cavity, the cartilaginous fish had no other way to pump water across their gill surfaces. Bony fish developed a way to do that using an **operculum**, a bony gill covering. The operculum not only protects the gills that lay underneath it, it is hinged so that its posterior edge can be pivoted away from the body. As it moves away from the body it pulls water from the buccal cavity out and across the gill surface. The movement of the operculum is coordinated with the opening and closing of the mouth and changes in the shape of the buccal cavity creating a steady flow of water across the gills.

Paired and unpaired fins with a fusiform body

The aquatic environment is a lot denser than the terrestrial, and animals that live in water face special challenges when they are trying to move. Compare running through a pool of waist-deep water to running along the deck beside the pool and you'll get the idea. The best way to move through a dense medium is to streamline the body, and that's what fish have done. Their bodies are **fusiform** with the tallest and widest part in the middle and the body tapering toward the anterior and posterior ends of the animal.

Unlike today's fish, the earliest fish didn't have any fins, and if they moved slowly, it probably wasn't a problem. As they moved faster, there was a potential for accidentally tilting, twisting, or turning and stability became a problem. The solution was fins, either unpaired or paired, which extended along the dorsal ventral line of the body and from the sides of the body. The unpaired fins include the **dorsal fin, anal fin,** and **caudal fin**. The **pectoral fins** in the front and **pelvic fins** behind them are both paired and attached to skeletal elements, **girdles**, that connect the right and left sides. But more than connecting the two sides, the girdles are an important site for attachment of the muscles that moved the fins.

✍ *What are the different types of caudal fins, and which types are found in the two classes of jawed fish?*

Scales

Vertebrate skin has the two layers: an outer **epidermis** and an underlying **dermis**. In fish, the epidermis is living and contains **mucous glands,** and the dermis underneath produces the **scales** that cover the body. Ancestrally the entire surface of the fish was covered in **dermal bone,** but in living fish species, scales have replaced this. Think of it as the change from solid metal armor that protected medieval warriors to the use of chain-link armor with its interconnected metal loops, although in the case of our fish, the scales are solid, not hollow. Some fish don't have scales or they're reduced to only certain parts of the body. In these fish, secretions of the mucous gland in the epidermal layer may help as a defense against predators.

A Closer Look Inside *Digital Zoology* - Jawed Fishes

Chondrichthyes: Dogfish shark

Sharks are cartilaginous fish and, as the name implies, their skeleton is made of **cartilage**. It's not clear if this is an ancestral or derived trait. The earliest fishes, including those identified as the ancestors to the sharks, the placoderms, were covered with bony dermal plates. This suggests that the presence of a purely cartilaginous skeleton is a derived trait. That may not be the whole story, though. The fossil record is biased for bony animals, not cartilaginous ones, and maybe there is a cartilaginous ancestor that has yet to be identified.

If you rub the skin of a shark, it feels like sandpaper, and dried shark skin has been used as sandpaper. Its texture results from **placoid scales**, **denticles**, that cover a shark's body. Like the scales of other fish, placoid scales are formed in the **dermal layer** of the skin, but they differ by piercing through the epidermis with a backwards projecting spine. The scale is made of **enamel**-covered **dentine,** and its large hollow base is embedded deep in the underlying dermis. The scales are important for reducing friction and drag as water flows over the surface of the body. Under ideal circumstances, friction and drag are at their lowest with laminar flow of the water. Surprisingly, perfectly smooth surfaces disrupt laminar flow, and a shark's scales direct the movement of water over the body to help maintain laminar as the shark swims. Competitive swimmers have picked up on this, and Speedo's "fastskin" full body bathing suit, which made its appearance at the Sydney Summer Olympic Games is designed on the same principles as shark skin. A shark's teeth are larger versions of these scales.

✍ *What is a shark's spiral valve, and what does it do?*

The urogenital system in the shark gives us a chance to see the close relationship between the reproductive and excretory systems and how they were organized in early vertebrates. The two systems are closely linked to each other, sharing the same ducts and tubes, and are embryological neighbors forming from adjacent tissues. Starting at the anterior end of the embryo, the **nephric ridge** differentiates forming tubules that will ultimately be **nephrons**. These are segmentally arranged down the length of the embryo and connected to each other by the **archinephric duct,** which drains to the outside. The entire length has the potential to be the kidney but kidneys form from parts of the ridge. Vertebrate kidneys form from either that anterior region, **pronephros**; middle region, **mesonephros**; or the posterior region, **metanephros**. The sharks urogenital system is an example of **opisthonephros,** a modified mesonephros extended posteriorly by the addition of more tubules.

✍ *What is the main metabolic waste in sharks, and how is it used for osmoregulation?*

Ancestrally, gonads developed from the tissues beside the kidney and shared the archinephric duct with the kidney to transport gametes outside the animal. This is still the case in the male shark, and the **testes** use the tubules at the anterior end of the kidney, no longer involved in excretion. Only the more posterior end of the male the kidney is excretory, and it dumps its wastes into a **urinary duct** leaving the original archinephric duct to be used by the reproductive system. The close association between the excretory and reproductive systems that occurs in males doesn't occur in females. Females have a separate tube, the **oviduct**, to transport their **eggs**. **Ovaries** shed their eggs into the coelom, and they are picked up by the ciliated funnel at the anterior end of the oviduct, the **oviducal funnel**. The eggs move down the oviduct by a combination of muscular contractions and cilia. The oviduct may have specialized regions for adding shell, **shell gland**, or other membranes to the egg. In the some sharks, such as the Dogfish shark, the posterior end of the oviduct is modified into an **ovisac** where the egg capsule can develop internally.

Osteichthyes: Perch

Bony fish have two different types of scales, **cycloid** and **ctenoid**. The two are similar in composition and differ only in their appearance. Cycloid scales are rounded with smooth edges; ctenoid scales have small projections or teeth that stick out from the epidermal covering. These small teeth may work the same way that the placoid scales do in sharks, enhancing the laminar flow of water over the body as the fish swims. There are concentric rings in both scales, and they are used to determine a fish's age.

Because living tissue is denser than water a fish would sink to the bottom if it didn't keep swimming or find some way to achieve the same buoyant density as the water that surrounds the fish. The bony fish achieve this with an air-filled **swim bladder**. It's not enough to have an internal bag of air. The amount of air in the bag has to change. As a fish swims deeper, the weight of the column of water above increases and compresses the air inside the swim bladder. More air will need to be pumped into the bladder to achieve neutral buoyancy at greater depths. It's the opposite as the fish rises to the surface; air will have to be removed from the bladder. Swim bladders evolved from the lungs of the early lungfish, and in today's bony fish, they are either connected to the alimentary tract by a pneumatic duct, **phystostomas,** or are disconnected, **physoclistous**. Phystostomas fish let air out through the pneumatic duct. Physoclistous fish can't do that. They have a highly vascularized surface of the bladder, the **ovale**, where gas moves from the bladder to the blood. Both types of fish add air to the bladder using the **gas gland** and its network of capillaries, the **rete mirable**, that secrete oxygen from the blood to the bladder.

 *How does a shark achieve a neutral buoyancy?*

Achieving neutral buoyancy was important and had a tremendous impact on the evolution of the bony fishes. A shark's **heterocercal** tail was combined with the pectoral and pelvic fins to constantly generate lift. If a shark stopped swimming, it would sink. With a swim bladder this doesn't happen in bony fish. Their tail is **homocercal** and the pectoral and pelvic fins can be used for something other than generating lift as they did in the shark. Bony fishes' fins are supported by dermal bone that forms **fin rays,** and musculature associated with the rays allows the fins to fold and unfold. This is best seen in the pectoral fin, which in bony fish is attached to the head. This fin bends, folds, and moves in a variety of ways that allow more precise and delicate movements associated with hovering.

The urogenital system of the bony fish has been modified from what we saw in the shark. Although they have an **opisthonephric kidney,** the excretory and reproductive systems are separate. In females, the **oviduct** and its **oviducal funnel** have been replaced with an **ovarian duct** that connects the ovary directly to the genital opening. Males have also developed a separate **testicular duct,** and the **archinephric duct** is used exclusively by the kidney.

Cross-References to Other McGraw-Hill Zoology Titles

Integrated Principles of Zoology, 11th edition. C.P. Hickman, L.S. Roberts & A. Larson. Chapter 26.

Animal Diversity, 2nd edition. C.P. Hickman, L.S. Roberts & A. Larson. Chapter 15.

Zoology, 5th edition, S.A. Miller & J.P. Harley. Chapter 18.

Vertebrate Biology, 1st edition. D.W. Linzey. Chapter 5.

Laboratory Studies in Integrated Principles of Zoology, 10th edition. C.P. Hickman, F. Hickman & L. Kats. Chapter 18.

General Zoology Laboratory Guide, 13th edition. C. Lytle. Chapter 16 & 17.

Structures Checklist

Here are some of the structures that you should be able to easily find in *Digital Zoology* and the specimens that you will be looking at in your lab. After reading your lab handout, you might want to add more and, depending on the equipment available in your lab, you might see more. As you study the material, you might also want to make some notes on how some of these structures looked or include a drawing in your lab notes.

Dogfish shark - external anatomy and skeltomuscular systems

External anatomy

- ☐ Abdominal pores
- ☐ Ampullae of Lorenzini
- ☐ Caudal fin
- ☐ Clasper (Male)
- ☐ Cloaca
- ☐ Dermis
- ☐ Dorsal fins (Anterior and Posterior)
- ☐ Endolymphatic pores
- ☐ Epidermis
- ☐ Eyes
- ☐ Fin spines
- ☐ Fin rays
- ☐ Gill slits
- ☐ Head
- ☐ Labial pouch and groove
- ☐ Lateral line
- ☐ Mouth
- ☐ Naris (Nostril)
- ☐ Pectoral fin
- ☐ Pelvic fin
- ☐ Placoid scales (Denticles)
- ☐ Rostrum
- ☐ Spiracle
- ☐ Trunk
- ☐ Urinary papilla (Female)
- ☐ Urogenital papilla (Male)
- ☐
- ☐
- ☐

Muscles

- ☐ 1st Levator (Palatoquadrate)
- ☐ Common coracoarcuals
- ☐ Coracobranchials
- ☐ Coracohyoid
- ☐ Coracomandibular
- ☐ Depressor of pectoral fin
- ☐ Dorsal constrictors
- ☐ Epaxial muscles
- ☐ Epibranchial musculature
- ☐ Horizontal septum
- ☐ Hypaxial muscles
- ☐ Interhyoid
- ☐ Levator hyomandibulae
- ☐ Levator of pectoral fin
- ☐ Linea alba
- ☐ Mandibular adductor
- ☐ Myomere
- ☐ Myosepta
- ☐ Radial muscles
- ☐ Spiracular muscle
- ☐ Ventral constrictors
- ☐
- ☐
- ☐
- ☐

Skeleton - Skull

- ☐ Antorbital process
- ☐ Antorbital shelf
- ☐ Basal plate
- ☐ Basibranchial cartilages
- ☐ Basihyal cartilage
- ☐ Ceratobranchial cartilages
- ☐ Ceratohyal cartilage
- ☐ Endolymphatic foramina
- ☐ Endolymphatic fossa
- ☐ Epibranchial cartilages
- ☐ Foramen magnum
- ☐ Hyomandibular cartilages
- ☐ Hypobranchial cartilages
- ☐ Infraorbital shelf
- ☐ Labial cartilage
- ☐ Meckel's cartilage
- ☐ Occipital condyles
- ☐ Olfactory capsules
- ☐ Palatopterygoquadrate cartilage
- ☐ Pharyngobranchial cartilages
- ☐ Postorbital process
- ☐ Postotic process
- ☐ Precerebral cavity
- ☐ Quadrate process
- ☐ Rostral fenestrae
- ☐ Rostral carina
- ☐ Rostrum
- ☐ Superficial ophthalmic foramina
- ☐ Supraorbital crest
- ☐ Supraotic crest
- ☐ Trochlear foramen
- ☐ Vagus foramina
- ☐
- ☐
- ☐

Skeleton - Trunk and appendages

- ☐ Acetabular surface
- ☐ Basal cartilages
- ☐ Caudal vertebrae
- ☐ Centrum
- ☐ Ceratotrichia
- ☐ Clasper cartilages (Male)
- ☐ Coracoid bar
- ☐ Glenoid surface
- ☐ Hemal canal
- ☐ Hemal spine
- ☐ Iliac process
- ☐ Intercalary plate
- ☐ Ischiopubic bar
- ☐ Mesopterygium
- ☐ Metapterygium
- ☐ Neural canal
- ☐ Neural spine
- ☐ Propterygium
- ☐ Radial cartilages
- ☐ Scapular cartilage
- ☐ Suprascapular cartilage
- ☐ Trunk vertebrae
- ☐
- ☐
- ☐

Dogfish shark - Internal anatomy

Digestive and related systems

- ☐ Bile duct
- ☐ Buccal cavity
- ☐ Cardiac region of stomach
- ☐ Cloaca
- ☐ Colon
- ☐ Duodenum
- ☐ Esophageal papillae
- ☐ Gall bladder
- ☐ Gastrosplenic ligament
- ☐ Hepatoduodenal ligament
- ☐ Hepatogastric ligament
- ☐ Ileum
- ☐ Lesser omentum
- ☐ Liver: right, medial, and left lobes
- ☐ Mesentery
- ☐ Mesorectum
- ☐ Pancreas, dorsal lobe
- ☐ Pancreas, ventral lobe
- ☐ Pyloric region of stomach
- ☐ Rectal gland
- ☐ Rectum
- ☐ Rugae of stomach
- ☐ Spiral valve
- ☐ Spleen
- ☐ Teeth
- ☐ Tongue
- ☐
- ☐
- ☐
- ☐
- ☐

Circulatory and respiratory systems

- ☐ Annular arteries
- ☐ Annular veins
- ☐ Anterior cardinal vein
- ☐ Atrium
- ☐ Branchial arteries, afferent
- ☐ Branchial arteries, efferent
- ☐ Caudal artery
- ☐ Caudal vein
- ☐ Celiac artery
- ☐ Common cardinal veins
- ☐ Conus arteriosus
- ☐ Coronary artery
- ☐ Demibranchs
- ☐ Dorsal aorta
- ☐ Esophageal artery
- ☐ Femoral vein
- ☐ Gastric artery
- ☐ Gastric vein
- ☐ Gastrohepatic artery
- ☐ Gastrosplenic artery
- ☐ Gill arch
- ☐ Gill filaments
- ☐ Gill rakers
- ☐ Gill rays
- ☐ Hepatic artery
- ☐ Hepatic portal vein
- ☐ Hepatic vein
- ☐ Holobranch
- ☐ Iliac arteries
- ☐ Iliac vein
- ☐ Inferior jugular vein
- ☐ Internal carotid artery
- ☐ Internal gill slits
- ☐ Intestinal arteries (Anterior and posterior)
- ☐ Intestinal veins (Anterior and Posterior)
- ☐ Lateral abdominal veins
- ☐ Lienomesenteric vein
- ☐ Mesenteric arteries
- ☐ Mesenteric veins
- ☐ Pancreaticomesenteric artery
- ☐ Pancreaticomesenteric vein
- ☐ Pericardial cavity
- ☐ Posterior cardinal vein
- ☐ Renal veins
- ☐ Sinus venosus
- ☐ Subclavian artery
- ☐ Subclavian vein
- ☐ Ventricle
- ☐
- ☐
- ☐
- ☐
- ☐

Urogenital system

- ☐ Cloaca
- ☐ Efferent ductules
- ☐ Kidneys
- ☐ Mesonephric ducts
- ☐ Mesorchium
- ☐ Mesotubarium
- ☐ Mesovarium
- ☐ Ostium tubae
- ☐ Ovaries
- ☐ Oviducts
- ☐ Seminal vesicle
- ☐ Shell gland
- ☐ Testes
- ☐ Urinary papilla
- ☐ Urogenital papilla
- ☐
- ☐
- ☐

Additional structures

- ☐
- ☐
- ☐
- ☐
- ☐
- ☐
- ☐
- ☐

Perch - External and internal anatomy

External anatomy

- ☐ Anal fin
- ☐ Anal opening
- ☐ Caudal fin
- ☐ Caudal peduncle
- ☐ Ctenoid scale
- ☐ Dermis
- ☐ Epidermis
- ☐ Excurrent naris (Nostril)
- ☐ Eye
- ☐ First dorsal fin
- ☐ Head
- ☐ Incurrent naris (Nostril)
- ☐ Lateral line
- ☐ Lips
- ☐ Lower mandible (Dentary)
- ☐ Maxilla
- ☐ Mouth
- ☐ Mucous glands
- ☐ Operculum
- ☐ Pectoral fin
- ☐ Pelvic fin
- ☐ Premaxilla
- ☐ Second dorsal fin
- ☐ Tail
- ☐ Trunk
- ☐ Urogenital pore
- ☐
- ☐
- ☐
- ☐

Muscular system

- ☐ Abductors of pectoral fin
- ☐ Abductors of pelvic fin
- ☐ Adductor caudalis
- ☐ Adductor mandibulae 1
- ☐ Adductor mandibulae 2
- ☐ Adductors of pelvic fin
- ☐ Dilator operculi
- ☐ Epaxial musculature
- ☐ Fin inclinator muscles
- ☐ Geniohyoideus posterior
- ☐ Geniohyoideus anterior
- ☐ Horizontal septum
- ☐ Hypaxial musculature
- ☐ Levator arcus palatini
- ☐ Levator hyoideus
- ☐ Levator operculi caudalis
- ☐ Levator operculi cranialis
- ☐ Myomeres
- ☐ Myosepta
- ☐ Rectus abdominis
- ☐ Sternohyoideus
- ☐ Trapezius
- ☐
- ☐
- ☐
- ☐

Skeletal system

- ☐ Articula
- ☐ Basibranchials
- ☐ Basipterygium
- ☐ Branchiostegl rays
- ☐ Caudal vertebrae
- ☐ Centrum
- ☐ Ceratobranchial
- ☐ Ceratohyal
- ☐ Cleithrum
- ☐ Coracoid
- ☐ Dentary
- ☐ Dentigerous plate
- ☐ Epibranchial
- ☐ Epipleural rib
- ☐ Frontal
- ☐ Gill supports
- ☐ Hemal arch
- ☐ Hyobranchial
- ☐ Hyomandibular
- ☐ Hypohyal
- ☐ Interopercular
- ☐ Lacrimal
- ☐ Lepidotrichia
- ☐ Maxilla
- ☐ Metapterygoid
- ☐ Nasal
- ☐ Neural spine
- ☐ Opercular
- ☐ Pharyngobranchial
- ☐ Pleural rib
- ☐ Premaxilla
- ☐ Preopercular
- ☐ Prootic
- ☐ Pterygiophores
- ☐ Quadrate
- ☐ Radials
- ☐ Scapula
- ☐ Sphenotic
- ☐ Subopercular
- ☐ Suborbital bones
- ☐ Supracleithrum
- ☐ Symplectic
- ☐ Trunk vertebrae
- ☐ Urohyal
- ☐
- ☐
- ☐
- ☐

Internal Anatomy

- ☐ Afferent arteries
- ☐ Anterior cardinal vien
- ☐ Atrium
- ☐ Branchial arch
- ☐ Bulbus arteriosus
- ☐ Common cardinal vein
- ☐ Duodenum
- ☐ Esophagus
- ☐ Gall bladder
- ☐ Gill filaments
- ☐ Gill rakers
- ☐ Head kidney
- ☐ Heart
- ☐ Intestine
- ☐ Kidney
- ☐ Liver
- ☐ Ovary
- ☐ Pericardial sack
- ☐ Postcardinal vein
- ☐ Pyloric ceca
- ☐ Sinus venosus
- ☐ Spleen
- ☐ Stomach
- ☐ Swim bladder
- ☐ Testes
- ☐ Urinary bladder
- ☐ Ventral aorta
- ☐ Ventricle
- ☐
- ☐
- ☐
- ☐

Crossword Puzzle - Jawed Fishes

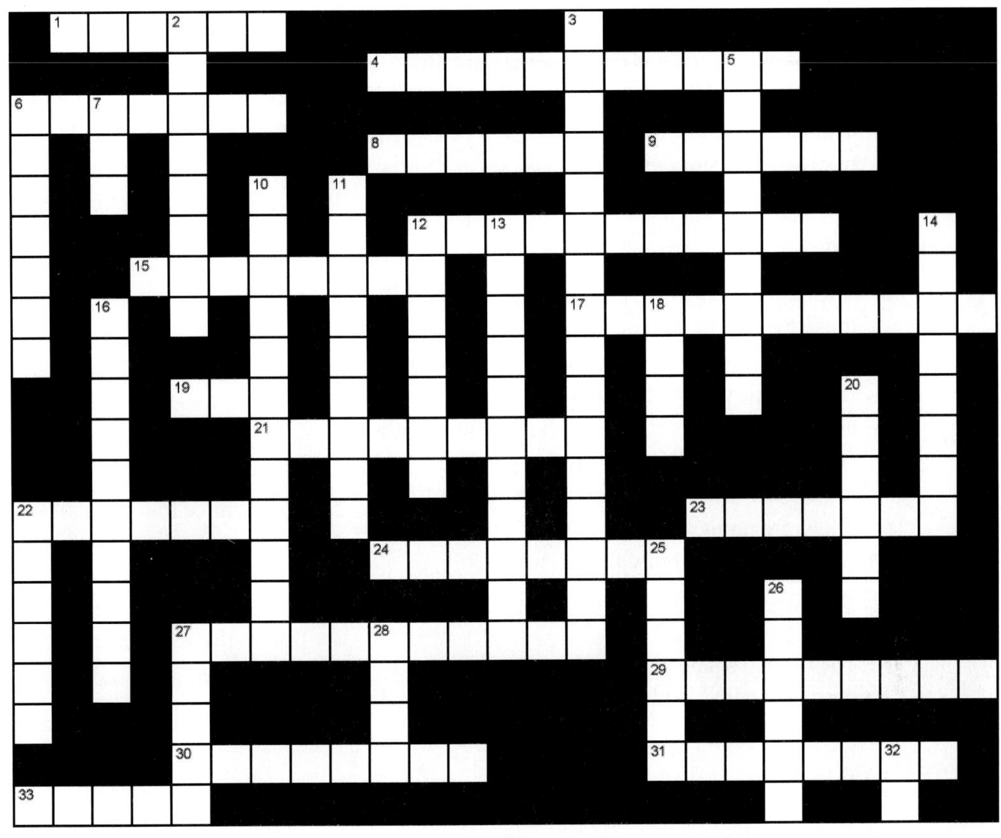

Puzzle solution. An interactive web based version of this puzzle, and its solution, are available on the *Digital Zoology* web site at www.mhhe.com/DigitalZoology/Students. With the interactive puzzle you can check to see if individual words or the whole puzzle is correct, and get hints for single letters.

Across

1 A shark's scales are formed from this layer of the integument. (6)

4 Bony fish maintain neutral density using this. (4,7)

6 These toothy scales are found in many bony fish. (7)

8 The posterior set of paired fins are these fins. (6)

9 Although a shark doesn't have a large stomach they do have a special valve in the intestine that slows the passage of food and increases the surface area of the gut. (6)

12 Although you'll find these in invertebrate, these blind-ended extensions of a bony fish's stomach are specialized sites for digestion. (11)

15 The most anterior pair of fins on the side of a shark are these fins. (8)

17 This "marvellous net" is where gas enters and leaves the swim bladder. (4,7)

19 Number of gill filaments on each gill. (3)

21 The skeleton of a shark is made of this material. (9)

22 In a male shark the pectoral fin is modified with one of these that's used in mating. (7)

23 The fins on the left side of the body are connected to those on the right by these. (7)

24 The body shape of a fish; it's the best one for moving through a dense medium such as water. (8)

27 These structures on the inside of the gill arch keep the food in the mouth and help to protect the delicate gills. (4,7)

29 This moveable flap covering a fish's gills helps to pump water over the gills. (9)

30 The platelike structures on the gill of a bony fish are where the gas exchange occurs. (8)

31 This oil helps keep a shark from sinking. (8)

33 The swim bladder ultimately becomes these in terrestrial vertebrates. (5)

Down

2 The name for the overlapping zigzag bands of muscle in the body wall of a fish. (8)

3 The ampullae of Lorenzini in a cartilaginous fish are this type of receptor. (16)

5 Unlike the scales of a shark, those in bony fish never pierce this layer of the integument; instead the layer covers the scales. (9)

6 There are no teeth on the edge of these almost circular scales in bony fish. (7)

7 A bony fish with neither pectoral nor pelvic fins. (3)

10 The unique shape of a sharks tail, which has a dorsal lobe that is much larger than the ventral lobe. (12)

11 You'll find these inside the dorsal lobe of a shark's tail. (9)

12 The type of scales found only in cartilaginous fish. (7)

13 This system detects vibrations in fish; it's the way that they hear. (11)

14 Another name for bony fish. (8)

16 In some fish a duct connects the swim bladder to this structure. (10)

18 Is this statement true or false?: Sharks lay eggs. (4)

20 A sharks teeth are highly modified versions of these surface structures. (6)

22 Another name for a tail fin is this type of fin. (6)

25 These glands are found in the epithelial layer of a bony fish's integument. (6)

26 Sharks have four types of fins that help stabilize the animal as it swims. This is one of the unpaired fins. (6)

27 After leaving the heart of a fish, blood flows first to these structures. (5)

28 The ventral unpaired fin in a shark. (4)

32 Can a shark breath through its nostril? (2)

Self Test - Jawed Fishes

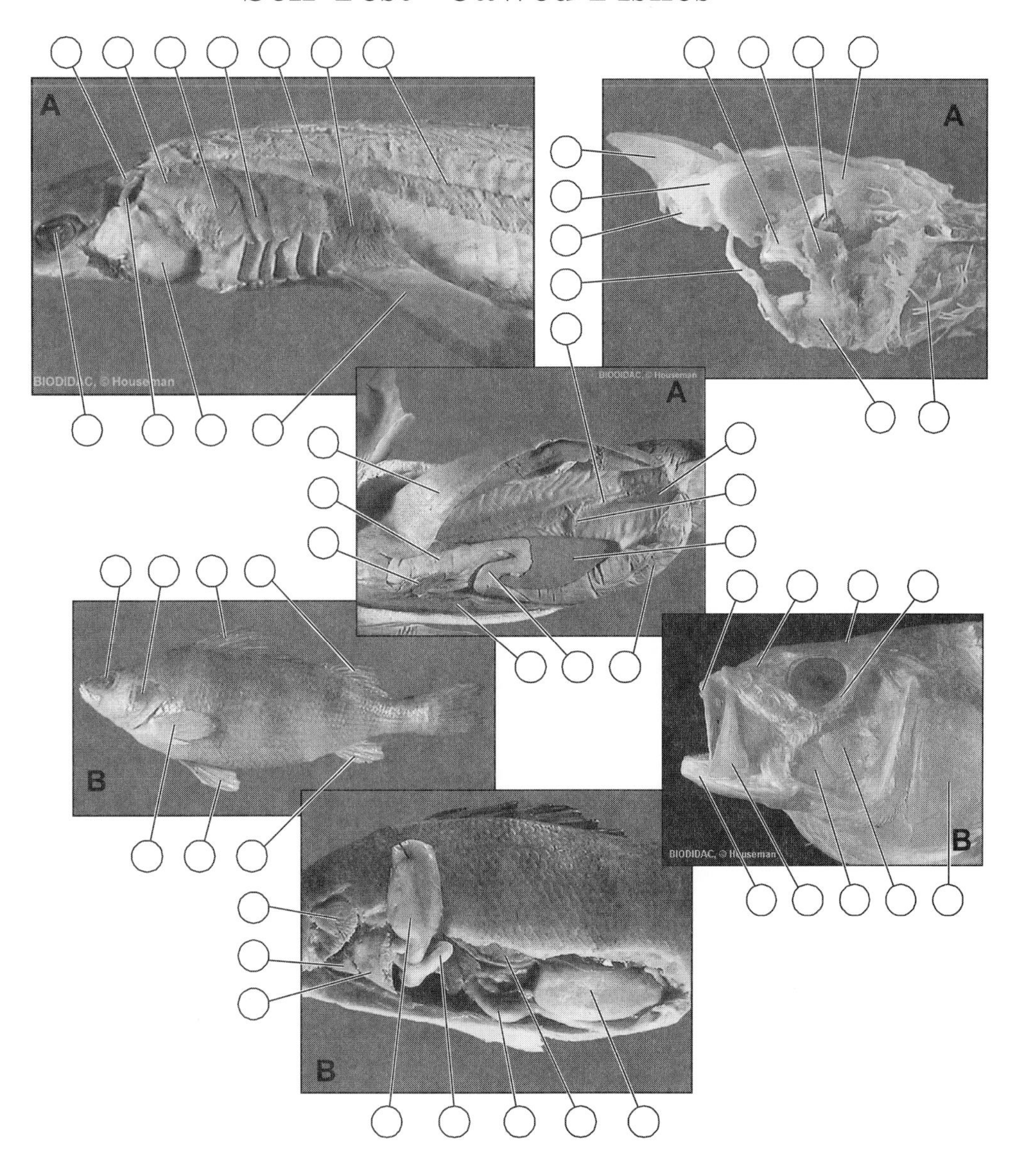

Use the following labels to identify the photographs. You may have to use a label more than once, and some labels may not be appropriate for the photographs. Answers are available on the *Digital Zoology* web site at www.mhhe.com/digitalzoology. Be sure to try the interactive Drag-and-Drop quizzes that are available on the *Digital Zoology* CD-ROM. A color version of this Self Test is available in the Adobe Acrobat version of the Student Workbook in the workbook folder on the *Digital Zoology* CD-ROM.

Specimens

A- Dogfish shark
B- Perch

Labels

1) Anal fin
2) Antorbital process
3) Cardiac region of stomach
4) Cucullaris
5) Dentary
6) Eye
7) First dorsal fin
8) Frontal
9) Gall bladder
10) Gill filaments
11) Gill rays
12) Heart
13) Horizontal septum
14) Infraorbital bones
15) Intestine
16) Kidney
17) Labial cartilage
18) Levator hyomandibulae

19) Levator of pectoral fin
20) Liver
21) Mandibular adductor
22) Maxilla
23) Meckel's cartilage
24) Mesenteric artery
25) Metapterygoid
26) Nasal bone
27) Olfactory capsule
28) Operculum
29) Optic pedicle
30) Palatoquadrate cartilage
31) Pectoral fin
32) Pelvic fin
33) Peripharyngeal band (Velum)
34) Pharynx
35) Postorbital process
36) Premaxilla
37) Proboscis coelom
38) Pyloric ceca
39) Pyloric region of stomach
40) Quadrate
41) Quadrate process
42) Rectal gland
43) Rostrum
44) Second dorsal constrictor
45) Second dorsal fin
46) Spiracle
47) Spiracular muscle
48) Spiral valve
49) Spleen
50) Stomach
51) Stomocord
52) Swim bladder
53) Testis
54) Third dorsal constrictor

AMPHIBIA

Inside *Digital Zoology*

As you explore the amphibians on the *Digital Zoology* CD-ROM, don't miss these learning tools:

Photos of dissected specimens of the mudpuppy and the frog including external anatomy and musculature, skeletal system, and internal anatomy.

Drag-and-drop quizzes on the external anatomy and musculature, skeletal system, and internal anatomy of the mudpuppy and the frog.

Interactive cladogram showing the major events that gave rise to vertebrate classes, and a summary of the key characteristics of each class are combined with an interactive glossary of terms.

Defining Differences

Some of the differences described in the following sections appear for the first time in the Amphibia, and they define the class. Others are important for understanding how these animals function. Whichever is the case, you'll want to watch for examples of these in the various amphibian specimens that you will be examining.

- Transition to land;
- tetrapod stance with a
- strong bony skeleton;
- double circuit circulatory system; and a
- moist, glandular skin.

Transition to land

Amphibians were the first vertebrates to move onto land. About 400 million years ago, during the Devonian Age, the earth was hot and dry. Aquatic environments often dried up or disappeared until flooding replenished them. As these bodies of water shrank, the density of fish and the other animals living in them increased. Crowding wasn't the only problem. The water became stagnant, depleted of oxygen. This was a problem for any animals that used **gills.** The solution was to breathe air, and that required **lungs**. **Gas bladders** in fish form from outpockets of the pharynx or gut and have two functions in modern fish – respiratory gas exchange or a mechanism for achieving neutral buoyancy. It's not clear which came first in the ancient fish but a fish with a gas bladder capable of gas exchange connected to the alimentary tract and external nostrils could poke its head above water and breathe. There's fossil evidence to suggest that is what happened.

That solves the problem of poor oxygen quality. What about the increasing density of organisms living in the diminishing body of water? The solution was to get up and leave, and the strong fins of the lobe-fined fishes helped them do that. Their fin musculature and bones were a modification for crawling across the bottom while they fed, and they also worked for moving on land. The first fish that wiggled up on land wasn't out to conquer this new environment; it was trying to get to a better body of water. Ultimately this was the beginning of Amphibia, the first vertebrates to spend a significant amount of time on land.

Earth's environment was to undergo another radical change after the dry Devonian. The Carboniferous was a time when the earth was wet and warm; ideal for the new amphibians, and they diversified quickly. Coming on land was not only advantageous for escaping predators and a deteriorating environment but this was the time when insects started to conquer the terrestrial environment. Amphibians found a food source on which no one else was feeding.

There was one thing that amphibians couldn't do on land, and that was reproduce. Their eggs still had to be laid in water and fertilized there. A swimming tadpole hatched from the egg and lived in the ancestral amphibian home until it was large enough to undergo **metamorphosis** and complete its life cycle on land.

Few species remain from the age of amphibians, and today's only representatives are salamanders, frogs, and a small group of caecilians. They represent the culmination of a long, diverse lineage of amphibians, and each is now specialized for its way of life. Neither of them looks like the supposed ancestral amphibian, and the same warning we made about cartilaginous and bony fish applies here. Don't think of salamanders as evolving into frogs. As we said with the two classes of fishes, it didn't happen that way.

Tetrapod stance

Amphibians were the first **tetrapods**. Their **appendicular skeleton** has undergone extensive modification, and now the **pectoral girdle** and **pelvic girdle** are firmly anchored to the **axial skeleton** using modified **vertebrae**. Attached to each of the girdles is a set of bones that form forelimbs and hindlimbs that hold the animal up on its four feet, the origins of the term tetrapod. This modification of the appendicular skeleton appears for the first time with the amphibians, and it's inherited by all the vertebrates that follow. The **taxon** is the Tetrapoda.

✍ *How many different types of vertebrae are there in vertebrates?*

Even though lung fish and lobe-finned fish may have had solutions to the problems of the Devonian, biologists still aren't certain which of the two was the one to come onto land. Whichever one it was, evolution of the tetrapod stance would follow the same sequence. The ancestral fish to the tetrapods strengthen its fins using **dermal bones,** called **fin rays**, anchored against a set of larger bones at their bases. The proposed changes required for the tetrapod stance saw a further strengthening of the bones that anchored the fin rays to the body and a decrease in the number of rays that ultimately became the **phalanges**. The first tetrapods probably had more than five phalanges on each limb but at some point the **pentatdactyl** plan appeared and all the vertebrates that followed inherited that plan.

✍ *In addition to the phalanges, what are the main bones of the tetrapod limbs?*

Strong bony skeleton

Movement on land required not only modifications of the pectoral and pelvic girdles, the **axial skeleton** had to change, too. As a skeletal element, the vertebral column of a fish had to resist compression along its length so that when the myomeres on each side contracted, the tail and trunk bent rather than shortened. Fish vertebrae are disc-shaped and embedded in the muscles of the tail and trunk. There was little chance of two vertebrae slipping by each other. That changed on land, and the vertebral column became a strengthening beam held up by the two sets of appendages. The appendages had to hook into the axial skeleton, which had to support the body. Vertebrae of tetrapods are considerably more complex than those of the fishes. Concave surfaces of the centrum fit into the convex surface of the adjacent vertebrae, and the neural arch is modified to form **zygapophyses** which extend forward and backwards.

Double circuit circulatory system

The amphibian heart is more complex than that of the fishes. Instead of one **atrium,** there are now two – one that receives blood from the lungs and the other, blood that has circulated through the body. We now have two circulatory routes through the body, the **pulmonary circuit** and the **systemic circuit,** and a **dual circuit circulatory system**. With only one **ventricle** the separation of the two circuits isn't perfect, and there is some mixing of the oxygenated blood from the pulmonary path with the deoxygenated blood from the systemic system. This is kept to a minimum by slight differences in the timing of when blood from each of the atria enter the ventricle, the walls of the ventricle, and the **spiral valve** in the **conus arteriosus**. It's long been thought that the amphibian heart, with its single ventricle, was an intermediate step on the way to the four-chambered heart seen in some reptiles and all birds and mammals. This may not be the case. The amphibian ventricle is perfectly adapted to an animal that may only occasionally use its lungs. There's no point sending blood to the lungs if they aren't always being ventilated. If there was a permanent division of the amphibian ventricle, this is exactly what would happen each time the heart would beat.

✍ *Blood oxygenated in the skin passes into which part of the circulatory system?*

Moist, glandular skin

Although lungs were important in the transition from an aquatic to a terrestrial existence, **cutaneous respiration**, gas exchange across the skin, is still important for amphibians. It's the reason they have a moist, glandular skin not protected by scales that would interfere with gas exchange. The skin is covered by a **stratum corneum** formed from a few layers of dead epidermal cells filled with **keratin**. This new epidermal component is thick enough to protect against abrasion but it's not so thick that it would interfere with respiration. For any animal that breathes through the skin there needs to be a mechanism to keep the skin moist. In amphibians, large **mucous glands** do that. These aren't the only glands in an amphibian's skin. Granular **poison glands** secrete their contents and help protect against predators.

A Closer Look Inside *Digital Zoology* - Amphibia

Necturus*: The mudpuppy*

Of the three amphibian orders the most primitive are the salamanders, amphibians with tails. *Necturus* may not look like any salamander that you've seen before because it lives its whole life in water. It is fishlike in how it swims and breathes and terrestrial with its tetrapod stance and skeletal modifications associated with that. This makes the mudpuppy a good animal for looking at the transition from water to land. As you examine your specimen, or the one in *Digital Zoology*, you might notice another unusual characteristic about this animal. The adult breathes with larval gills.

The mudpuppy is an example of **neoteny**. It's a special case of **paedomorphosis**, which describes animals that retain their larval appearance but are sexually mature. A larval tunicate that underwent paedomorphosis is how Garstang proposed that the first chordates evolved. He didn't dream this up. There are biological precedents that he drew on, and the mudpuppy is one of them. In some ways the mudpuppy is an overgrown tadpole with legs and a reproductive system.

A mudpuppy's skeleton is composed of both **cartilaginous** and **bony** elements. The skull retains elements of the cartilaginous **chondrocranium,** which becomes covered by the dermal bones of the **dermatochranium** as the animal matures. A fish's head doesn't articulate with the vertebral column, so fish don't have necks. The mudpuppy doesn't either. Even though mudpuppies aren't terrestrial, their **vertebrae** have the overlapping processes, **prezagopophysis** and **postzygapophysis,** which strengthen the vertebral column so that it can be supported by the forelimb and hind limb. The limbs aren't very big in a mudpuppy, and the **pectoral girdle** and **pelvic girdle** are a combination of cartilaginous and bony elements. Like cartilaginous fish, the pectoral girdle isn't attached to the vertebral column but is instead anchored in muscle. Only the pelvic girdle has **ilia** that connect it to the vertebral column as it does in terrestrial tetrapods

The musculature is another good example of the transition to land with each girdle having its own set of muscles, which in salamanders are used to lift the body up and off the ground and to allow it to walk. But, like a fish, the mudpuppies axial musculature is still extensive with **epaxial** muscles positioned above a **horizontal septum** and **hypaxial** muscles below. The hypaxial musculature is the most extensive, and it's easy to see the three overlapping sets of **oblique muscles** below the horizontal septum.

Internally the urogenital system is fishlike and closely resembles that of the sharks. In the male, that anterior part of the **opisthonephric kidney** is involved in transporting sperm from the **testes** to the **archinephic duct** (mesonephric duct). The female urogenital system is also like a shark's and eggs released from the **ovary** are swept into an **oviducal funnel** and the anterior of the **oviduct.**

Rana*: The grass frog*

The frog is a common dissection in zoology courses. It gives the opportunity to see some of the ancestral amphibian characters but also to see how the amphibian body has been modified for **saltatory locomotion**, jumping. The explosive muscular forces required for jumping have resulted in extensive modifications of both the axial and appendicular skeletons.

In jumping animals, the part of the leg that touches the ground is usually enlarged so that the forces generated from pushing up and off the ground can be spread over as much surface as possible. Frogs are no different, and their long toes help to increase the size of the hindfoot. Another modification that increases the size of the hindfoot are the two tarsal bones that have elongated to become the **astragulus (tibiale)** and **calcaneum (fubulare)**. A quick glance at the skeleton of a frog and you might mistake these two modified tarsal bones as the **tibia** and **fibula** because of their paired appearance. The real tibia and fibula have fused to form the **tibiofibula** bone, another modification that strengthens the hindlimb. In vertebrates, the **ilium** connects the pelvic girdle to the **sacral vertebrae,** and in frogs its length increases the size of the hindleg. There was an added advantage to this large foot that had nothing to do with jumping and that was for **natatorial locomotion**, swimming.

The axial skeleton has also been modified for jumping, and the **urostyle** is formed from the fusion of a number of postsacral vertebrae. The urostyle is positioned between the two large ilial bones. Muscles between the two help to strengthen the connection between the appendicular and axial skeletons. Even the forelimbs are modified for jumping, or more precisely, landing. When a frog jumps, it lands with its front legs first. The radius and ulna have fused into a stronger **radioulna**.

Cross-References to Other McGraw-Hill Zoology Titles

Integrated Principles of Zoology, 11th edition. C.P. Hickman, L.S. Roberts & A. Larson. Chapter 27.

Animal Diversity, 2nd edition. C.P. Hickman, L.S. Roberts & A. Larson. Chapter 16.

Zoology, 5th edition. S.A. Miller & J.P. Harley. Chapter 19.

Laboratory Studies in Integrated Principles of Zoology, 10th edition. C.P. Hickman, F. Hickman & L. Kats. Chapter 19.

General Zoology Laboratory, 13 edition. C. Lytle. Chapter 18.

Vertebrate Biology, 1st edition. D.W. Linzey. Chapter 6.

Structures Checklist

Here are some of the structures that you should be able to easily find in *Digital Zoology* and the specimens that you will be looking at in your lab. After reading your lab handout, you might want to add more and, depending on the equipment available in your lab, you might see more. As you study the material, you might also want to make some notes on how some of these structures looked or include a drawing in your lab notes. (Structures indicated by * may be hard to see.)

Mudpuppy - External anatomy and skeletomuscular systems

External anatomy

- ☐ Chromatophores
- ☐ Cloacal opening
- ☐ Dermis
- ☐ Epidermis
- ☐ External gills
- ☐ External naris (Nostrils)
- ☐ Eyes
- ☐ Gular fold
- ☐ Head
- ☐ Lips
- ☐ Mouth
- ☐ Mucous glands
- ☐ Pectoral limb
- ☐ Pelvic limb
- ☐ Poison glands
- ☐ Tail
- ☐ Trunk
- ☐
- ☐
- ☐
- ☐
- ☐

Musculature - Head and trunk

- ☐ Anterior mandibular levator
- ☐ Branchial levators
- ☐ Branchiohyoid
- ☐ Dorsalis trunci
- ☐ Dorsalis trunci
- ☐ Epaxial musculature
- ☐ Eternal oblique
- ☐ External mandibular levator
- ☐ Geniohyoid
- ☐ Horizontal septum
- ☐ Hypaxial musculature
- ☐ Interhyoid
- ☐ Intermandibular
- ☐ Internal oblique
- ☐ Linea alba
- ☐ Mandibular depressor
- ☐ Myomeres
- ☐ Myosepta
- ☐ Rectus abdominus
- ☐ Rectus cervicis
- ☐ Transversus abdominus
- ☐
- ☐
- ☐

Musculature - Appendages

- ☐ Caudofemoralis
- ☐ Caudopuboischiotibilias (Caudocruralis)
- ☐ Coracobrachialis
- ☐ Cucullaris
- ☐ Dorsal scapulae
- ☐ Humeroantebrachialis
- ☐ Ilioextensprius
- ☐ Iliofubularis
- ☐ Iliotibialis
- ☐ Ishioflexorius
- ☐ Latissimus dorsi
- ☐ Pectoralis
- ☐ Pectoriscapularis
- ☐ Procoracohumeralis
- ☐ Pubioischiofemoralis internus
- ☐ Puboischiofemoralis externus
- ☐ Puboischiotibialis
- ☐ Pubotibialis
- ☐ Supracotacoideus
- ☐ Triceps
- ☐
- ☐
- ☐
- ☐

Skeleton - Skull and trunk

- ☐ Angula
- ☐ Basibranchials
- ☐ Capitulum
- ☐ Caudal vertebrae
- ☐ Centrum
- ☐ Ceratobranchials
- ☐ Ceratohyal
- ☐ Cervical vertebra (atlas)
- ☐ Columella
- ☐ Dentray
- ☐ Diapophysis
- ☐ Epibranchials
- ☐ Ethmoid
- ☐ Foramen magnum
- ☐ Frontals
- ☐ Hyoid arch
- ☐ Hypohyal
- ☐ Meckel's cartilage
- ☐ Neural arch
- ☐ Neural canal
- ☐ Neural spine
- ☐ Occiptal condyle
- ☐ Opisthotic
- ☐ Palatopterygoids
- ☐ Parapophysis
- ☐ Parietals
- ☐ Postzygapophysis
- ☐ Premaxilla
- ☐ Prezygapophysis
- ☐ Prootic
- ☐ Quadrates
- ☐ Rib
- ☐ Sacral vertebra
- ☐ Splenial
- ☐ Squamosals
- ☐ Transverse process
- ☐ Trunk vertebrae
- ☐ Tuberculum
- ☐ Visceral arches
- ☐ Vomers
- ☐
- ☐
- ☐
- ☐
- ☐

Mudpuppy - Skeletal (cont.), and internal systems

Skeleton - appendages

- ☐ Acetabulum
- ☐ Carpals
- ☐ Coracoid cartilage
- ☐ Femur
- ☐ Fibula
- ☐ Glenoid fossa
- ☐ Humerus
- ☐ Ilium
- ☐ Ischium
- ☐ Ishia
- ☐ Metacarpals
- ☐ Phalanges
- ☐ Procoracoid cartilage
- ☐ Pubis
- ☐ Puboishiadic plate
- ☐ Radius
- ☐ Scapula
- ☐ Suprascapular cartilage
- ☐ Tibia
- ☐ Ulna
- ☐
- ☐
- ☐
- ☐
- ☐
- ☐

Digestive and related systems

- ☐ Dentary teeth
- ☐ Duodenum
- ☐ Esophagus
- ☐ Glottis
- ☐ Ileum
- ☐ Liver
- ☐ Mouth
- ☐ Pancreas
- ☐ Pharynx
- ☐ Premazillary teeth
- ☐ Pterygoid teeth
- ☐ Rectum
- ☐ Spleen
- ☐ Splenial teeth
- ☐ Stomach
- ☐ Vomerine teeth
- ☐
- ☐
- ☐
- ☐

Circulatory and respiratory systems

- ☐ Afferent branchial arteries
- ☐ Anterior mesenteric artery
- ☐ Atria
- ☐ Axillary vein
- ☐ Brachial vein
- ☐ Bulbus arteriousus
- ☐ Coacal arteries
- ☐ Celiac artery
- ☐ Celiacomesenteric artery
- ☐ Common cardinal veins
- ☐ Conus arteriosus
- ☐ Cutaneous vein
- ☐ Dorsal aorta
- ☐ Efferent branchial arteries
- ☐ External carotid artery
- ☐ External jugular vein
- ☐ Gastric artery
- ☐ Gastrosplenic vein
- ☐ Genital arteries
- ☐ Hepatic artery
- ☐ Hepatic portal vein
- ☐ Iliac arteries
- ☐ Internal carotid artery
- ☐ Internal jugular vein
- ☐ Intestinal arteries
- ☐ Lung
- ☐ Mesenteric vein
- ☐ Pancreaticoduodenal artery
- ☐ Posterior cardinal vein
- ☐ Posterior mesenteric artery
- ☐ Posterior vena cava
- ☐ Pulmonary artery
- ☐ Pulmonary vein
- ☐ Radix aorta
- ☐ Renal arteries
- ☐ Sinus venosus
- ☐ Splanic artery
- ☐ Subclavian artery
- ☐ Subclavian vein
- ☐ Ventral abdominal vein
- ☐ Ventricle
- ☐ Vertebral artery
- ☐
- ☐
- ☐
- ☐
- ☐
- ☐
- ☐
- ☐

Urogenital systems

- ☐ Cloaca
- ☐ Efferent ductules
- ☐ Kidneys
- ☐ Mesonephric ducts (Archinephric duct)
- ☐ Mesorchium
- ☐ Mesotubarium
- ☐ Mesovarium
- ☐ Ostium (Oviduct)
- ☐ Ovaries
- ☐ Testes
- ☐ Urinary bladder
- ☐
- ☐
- ☐
- ☐
- ☐
- ☐

Additional structures

- ☐
- ☐
- ☐
- ☐
- ☐
- ☐
- ☐
- ☐
- ☐
- ☐
- ☐
- ☐

Frog - Extrenal anatomy and skeletomuscular systems

External anatomy

- ☐ Ankle
- ☐ Antebrachium
- ☐ Brachium
- ☐ Chromophores
- ☐ Cloacal opening
- ☐ Crus (Leg)
- ☐ Dermis
- ☐ Digits (Fingers)
- ☐ Elbow
- ☐ Epidermis
- ☐ External nares (Nostrils)
- ☐ Eye
- ☐ Eye lids
- ☐ Femur (Thigh)
- ☐ Head
- ☐ Knee
- ☐ Manus (Hand)
- ☐ Mouth
- ☐ Mucous glands
- ☐ Nictitating membrane
- ☐ Pes (Foot)
- ☐ Poison glands
- ☐ Pollex (Thumb)
- ☐ Tibiotarsalus
- ☐ Trunk
- ☐ Tympanic membrane
- ☐
- ☐
- ☐
- ☐

Musculature - Head and trunk

- ☐ Coccygeoiliacus
- ☐ Cutaneous abdominis
- ☐ Cutaneous pectoris
- ☐ Depressor mandibulae
- ☐ Dorsalis scapulae
- ☐ External oblique
- ☐ Geniohyoideus
- ☐ Iliolumbaris
- ☐ Internal oblique
- ☐ Latissimus dorsi
- ☐ Longissimus dorsi
- ☐ Masseter
- ☐ Mylohyoid (Submaxillary)
- ☐ Omohyoideus
- ☐ Petrohyoideus
- ☐ Pterygoideus
- ☐ Rectus abdominis
- ☐ Rhomboideus
- ☐ Sternohyoideus
- ☐ Subhyoid
- ☐ Temporalis
- ☐ Transverse oblique
- ☐
- ☐
- ☐
- ☐

Musculature - Appendages

- ☐ Abductor indicis longus
- ☐ Adductor longus
- ☐ Adductor magnas
- ☐ Coracobranchialis
- ☐ Crualis (Vastus internus)
- ☐ Cucullaris
- ☐ Deltoideus
- ☐ Dorsalis scapulae
- ☐ Extensor carpi radialis
- ☐ Extensor carpi ulnaris
- ☐ Extensor cruris brevis
- ☐ Extensor digitorum communis longus
- ☐ Flexor carpi radialis
- ☐ Flexor carpi ulnaris
- ☐ Flexor digitorum
- ☐ Gastrocnemius
- ☐ Gluteus (Iliacus externus)
- ☐ Gluteus magnus (Vastus externus)
- ☐ Gracilis major (Rectus internus major)
- ☐ Gracilis minor (Rectus internus minor)
- ☐ Iliacus internus
- ☐ Iliofemorallis
- ☐ Iliofubularis (Biceps femoris)
- ☐ Latissimus dorsi
- ☐ Palmaris longus
- ☐ Pectineus
- ☐ Pectoralis major
- ☐ Pectoralis minor
- ☐ Peroneus
- ☐ Plantaris longus
- ☐ Quadratus femoris
- ☐ Sartorius
- ☐ Semimembranosus
- ☐ Semitendinosus
- ☐ Tarsalis anterior
- ☐ Tarsalis posterior
- ☐ Tendon of Achilles
- ☐ Tensor fasciae latae
- ☐ Tibialis anticus longus
- ☐ Tibialis posticus longus
- ☐ Triceps brachii (Anconeus)
- ☐ Triceps femoris
- ☐
- ☐
- ☐
- ☐

Skeleton - Skull and trunk

- ☐ Articular process (Zygopophyses)
- ☐ Atlas
- ☐ Centrum
- ☐ Columella
- ☐ Dentary
- ☐ Exoccipital
- ☐ Foramen magnum
- ☐ Frontoparietal
- ☐ Maxilla
- ☐ Nasal
- ☐ Neural arch
- ☐ Neural spine
- ☐ Occiptal condyle
- ☐ Orbit
- ☐ Otic capsule
- ☐ Palatine
- ☐ Parasphenoid
- ☐ Premaxilla
- ☐ Prootic
- ☐ Pterygoid
- ☐ Quadrate
- ☐ Quadratojugal
- ☐ Sacral vertebrae
- ☐ Sphenethmoid
- ☐ Squamosal
- ☐ Sternum
- ☐ Supraoccipital
- ☐ Transverse process
- ☐ Trunk vertebrae
- ☐ Urostyle
- ☐ Vomer
- ☐
- ☐
- ☐
- ☐

Frog - Skeletal (cont.), and internal systems

Skeleton - appendages

- ☐ Acetabulum
- ☐ Astragalus (Tarsus)
- ☐ Calcaneus (Tarsus)
- ☐ Carpals
- ☐ Clavicle
- ☐ Coracoid
- ☐ Femur
- ☐ Glenoid fossa
- ☐ Humerus
- ☐ Ilium
- ☐ Ischium
- ☐ Metacarpals
- ☐ Metatarsals
- ☐ Phalanges
- ☐ Pubis
- ☐ Radioulna
- ☐ Scapula
- ☐ Sternum
- ☐ Suprascapula
- ☐ Tarsals
- ☐ Tubiofibula
- ☐
- ☐
- ☐
- ☐
- ☐

Digestive and related systems

- ☐ Pharynx
- ☐ Esophagus
- ☐ Eustachian tube
- ☐ Tongue
- ☐ Vomerine teeth
- ☐ Internal nares
- ☐ Glottis
- ☐ Vocal sacs
- ☐ Stomach
- ☐ Pyloric sphincter
- ☐ Duodenum
- ☐ Jejunoileum
- ☐ Pancreas
- ☐ Cystic duct
- ☐ Common bile duct
- ☐ Colon
- ☐ Cloaca
- ☐ Liver
- ☐ Gall bladder
- ☐ Spleen
- ☐ Mesenteries
- ☐
- ☐
- ☐
- ☐

Circulatory and respiratory systems

- ☐ Anterior mesenteric artery
- ☐ Anterior vena cava
- ☐ Brachial artery
- ☐ Brachial vein
- ☐ Celiacomesenteric artery
- ☐ Celic artery
- ☐ Common carotid artery
- ☐ Conus arteriosus
- ☐ Cutaneous artery
- ☐ Dorsal aorta
- ☐ Dorsolumbar vein
- ☐ External carotid artery
- ☐ External jugular
- ☐ Femoral artery
- ☐ Femoral vein
- ☐ Gastric artery
- ☐ Gastric vein
- ☐ Gonadial vein
- ☐ Heart
- ☐ Hepatic artery
- ☐ Hepatic portal vein
- ☐ Hepatic vein
- ☐ Iliac artery
- ☐ Iliac vein
- ☐ Innominate vein
- ☐ Internal carotid artery
- ☐ Internal jugular vein
- ☐ Left atrium
- ☐ Lingual vein
- ☐ Lumbar arteries
- ☐ Lung
- ☐ Mandibular vein
- ☐ Mesenteric veins
- ☐ Musculocutaneous vein
- ☐ Pelvic vein
- ☐ Posterior mesenteric artery
- ☐ Posterior vena cava
- ☐ Pulmocutaneous arch
- ☐ Pulmonary artery
- ☐ Pulmonary vein
- ☐ Renal artery
- ☐ Renal portal vein
- ☐ Renal vein
- ☐ Right atrium
- ☐ Sciatic artery
- ☐ Sciatic vein
- ☐ Sinus venosus
- ☐ Splenic artery
- ☐ Splenic vein
- ☐ Subclavian artery
- ☐ Subclavian vein
- ☐ Subscapular vein
- ☐ Systemic arch
- ☐ Truncus arteriosus
- ☐ Urogenital arteries
- ☐ Ventral abdominal vein
- ☐ Ventricle
- ☐
- ☐
- ☐
- ☐
- ☐

Urogenital systems

- ☐ Adrenal glands
- ☐ Archinephric duct (Wolffian duct)
- ☐ Efferent ductules
- ☐ Fat body
- ☐ Kidney
- ☐ Mesorchium
- ☐ Mesovarium
- ☐ Ostium
- ☐ Oviduct
- ☐ Seminal vesicle
- ☐ Testes
- ☐ Ureters
- ☐ Urinary bladder
- ☐ Uterus
- ☐
- ☐
- ☐
- ☐
- ☐

Additional structures

- ☐ Subcutaneous lymph sacs
- ☐
- ☐
- ☐
- ☐
- ☐

Crossword Puzzle - Amphibia

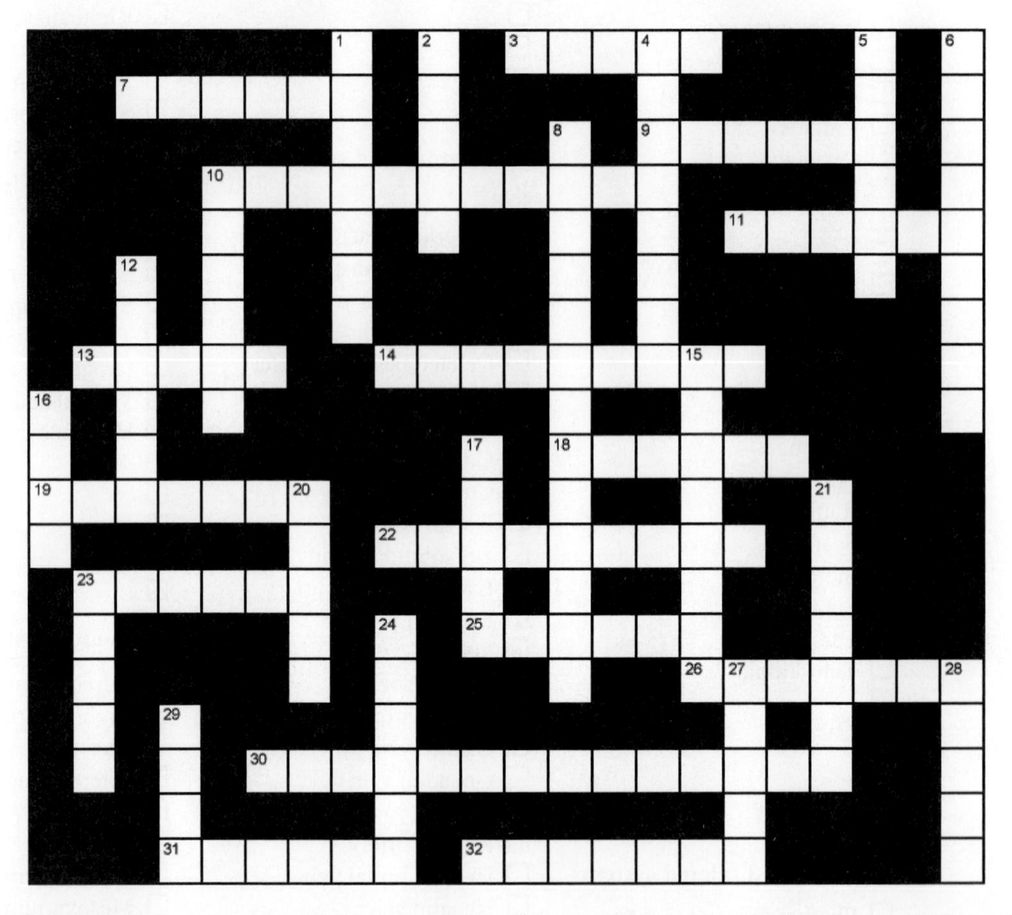

Puzzle solution. An interactive web based version of this puzzle, and its solution, are available on the *Digital Zoology* web site at www.mhhe.com/DigitalZoology/Students. With the interactive puzzle you can check to see if individual words or the whole puzzle is correct, and get hints for single letters.

Across

3 The single upper bone in the tetrapod hind limb. (5)

7 Describes a circulatory system with two routes that blood can take as it passes through the heart. (6)

9 Compared to water, there is more of this in air. (6)

10 Frogs, caecillans, and these are the only living amphibians. (11)

11 This and the tibia form the lower bones of the tetrapod hind limb. (6)

13 To survive, all amphibians have to lay their eggs in this. (5)

14 The transition from water to air appears in the amphibian's life cycle, ontogeny, as well as this history of the group. (9)

18 These glands in an amphibian are defensive. (6)

19 Probable food for the first amphibians on land. (7)

22 The extra loop in the amphibian circulatory system missing in fish. (9)

23 The amphibian voice box. (6)

25 This valve in the ventricle of an amphibian heart separates blood flow from the parts of the circulatory system. (6)

26 This bone, an addition to the axial skeleton of a frog, helps support and contain the internal organs. (7)

30 These gives an amphibian's skin its colors. (14)

31 This and the ulna form the lower bones of the tetrapod forearm. (6)

32 An immature larval amphibian. (7)

Down

1 Outer cells of the amphibian epidermis have this to strengthen them, but not so much that gas exchange is inhibited. (7)

2 With the movement on land, this type of musculature isn't as important in amphibians as it was in fishes. (5)

4 In a frog these are formed from fused vertebrae and are an adaptation related to jumping. (8)

5 A ligular flap refers to this structure in a frog. (6)

6 The presence of fore and hind limbs identifies amphibians and all the vertebrates to follow as this group of animals. (9)

8 The transition from a tadpole to an adult frog. (13)

10 Unlike fish, you won't find these on the surface of an amphibian. (6)

12 To be able to keep amphibian skin moist you'll find many of these in the skin. (6)

15 An amphibian can breathe through these when its mouth is shut. (8)

16 The main respiratory surface in most amphibians. (4)

17 Many amphibians demonstrate paedomorphosis and these appear in the adult. (5)

20 In amphibians blood flow to the lungs comes from this aortic arch. (5)

21 The single upper bone in the tetrapod forelimb. (7)

23 Limbs were one of the things advantageous for moving on land; this was one of the others. (5)

24 These glands in an amphibian help to keep its skin moist. (6)

27 Number of chambers in an amphibian heart. (5)

28 Lungs, skin, and the inner surface of this can be used to varying degrees in amphibians for gas exchange. (5)

29 The number of limbs that tetrapod refers to. (4)

Self Test - Amphibia

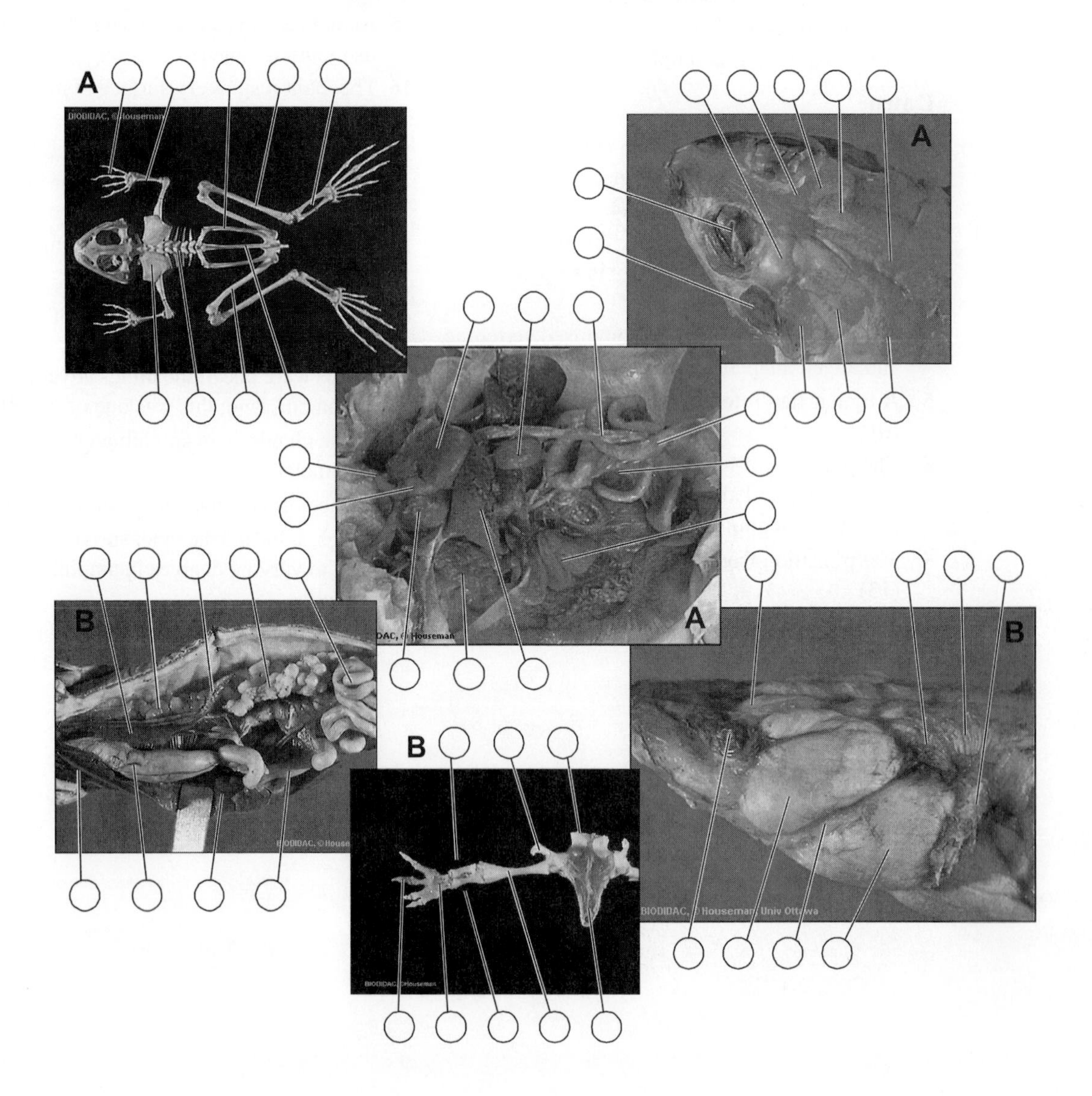

Use the following labels to identify the photographs. You may have to use a label more than once, and some labels may not be appropriate for the photographs. Answers are available on the *Digital Zoology* web site at www.mhhe.com/digitalzoology. Be sure to try the interactive Drag-and-Drop quizzes that are available on the *Digital Zoology* CD-ROM. A color version of this Self Test is available in the Adobe Acrobat version of the Student Workbook in the workbook folder on the *Digital Zoology* CD-ROM.

Specimens

A- *Rana* - Frog
B- *Necturus* - Mudpuppy

Labels

1) Conus arteriosus
2) Cucullaris
3) Depressor mandibulae
4) Depressor mandibulae
5) Dilator laryngis
6) Dorsalis scapulae
7) Duodenum
8) Eye
9) Femur
10) Fibula
11) Gall bladder
12) Gill
13) Hepatic vein
14) Ileum
15) Ilium
16) Interhyoideus

17) Ischium
18) Latissimus dorsi
19) Levatores arcuum
20) Levator mandibulae anterior
21) Levator mandibulae externus
22) Liver
23) Lung
24) Mesentery
25) Metatarsals
26) Ovary
27) Oviduct
28) Phalange
29) Pterygoideus
30) Pubis
31) Radioulna
32) Rhomboideus
33) Right atrium of heart
34) Small intestine
35) Spleen
36) Stomach
37) Suprascapula
38) Tarsus calcaneum
39) Temporalis
40) Tibia
41) Tibiofibula
42) Truncus arteriosus
43) Tympanum
44) Urostyle
45) Ventral abdominal vein
46) Ventricle of heart
47) Vertebrae

MAMMALIA

Inside *Digital Zoology*

As you explore the mammals on the *Digital Zoology* CD-ROM, don't miss these learning tools:

Photos of dissected specimens of the fetal pig including external anatomy and musculature, skeletal system, and internal anatomy.

Drag-and-drop quizzes on the external anatomy and musculature, skeleton, and internal anatomy of the fetal pig.

Interactive cladogram showing the major events that gave rise to vertebrate classes, and a summary of the key characteristics of each class are combined with an interactive glossary of terms.

Defining Differences

Some of the differences described in the following sections appear for the first time in the Mammalia, and they define the class. Others are important for understanding how these animals function. Whichever is the case, you'll want to watch for examples of these in the specimens that you will be examining.

- Homeothermic endotherms with
- hair,
- glandular skin,
- heterodent diphyodont dentition, a
- muscular diaphragm, and
- extended care of young.

Homeothermic endotherms

Mammals and birds are the only two taxa of animals that maintain their bodies at a precise and constant temperature, **homeothermy,** and they generate heat to maintain that temperature, **endothermy**. All of life's processes are based on biochemical reactions, and reaction rates are temperature dependent. The lower the temperature, the slower the reaction. Raise the temperature and the reactions speed up only to a point that the enzymes start to breakdown from the heat. There is an advantage to guaranteeing that life's biochemical reactions happen at the same temperature, and homeothermic endotherms have this advantage.

For an animal to maintain a constant body temperature there needs to be a way to generate heat when it's necessary and a way to get rid of it if the animal gets too hot. Heat is a normal by-product of the biochemical reactions of metabolism and is used by mammals and birds as a source of heat. **Shivering** is muscle contractions producing metabolic heat as a by-product of the contractions. Oxidative metabolism of brown fats is another source of metabolic heat. A homeotherm can't spend its time generating heat, and there's an advantage to retaining heat by insulating the body. The fur coat is a mammal's insulative covering.

 *How does countercurrent blood flow in the extremities minimize cooling of the body core of a mammal?*

That takes care of generating heat; what about cooling? A mammal's temperature is lowered by evaporation of watery sweat poured onto the surface of the skin by sweat glands. Sweat glands aren't the only way that mammals use evaporative cooling. The nasal passages or tongue are other important surfaces that mammals may use to lower their body temperature.

Hair

Just as feathers are unique to birds, **hair** is only found on mammals. Mammalian skin has the same two layers found in all vertebrates, an outer **epidermis** and an underlying **dermis**. The epidermis is much thicker than the dermis and is differentiated into a number of layers. New epidermal cells form in the bottom **stratum germinativum,** and as newer cells form, they push the older cells above them to the surface. As this happens, cells become **keratinized,** and the outermost layer of the skin, **stratum corneum,** is formed from dead keratinized cells, which waterproof mammalian skin and protect it against abrasion.

The **hair follicle** is a fold in the lower layers of the epidermis that penetrates deep into the dermal part of the skin. The dermis forms the **dermal papillae** at the base of the hair, and it's surrounded by the **epidermal hair matrix** formed from the stratum germinativum. It's the matrix that produces the hair. Mitotic cell divisions add new living cells at the hairs' base. As the new cells are added, the old keratinized cells of the hair die and are pushed up and out of the follicle. The result is a hair shaft composed of a **central cortex** surrounded by a **medula** completely wrapped inside overlapping **scale cells**.

Most mammals are covered with **fur**, or **pelage**, formed from two types of hairs. Longer, tough **guard hairs** protect the softer insulating **underfur** found next to the body. Guard hairs can also be modified into **quills** in animals such as the porcupine, or tactile **vibrissae**, whiskers found around the snouts of many mammals. Hairs usually lie at an angle to the skin and point backward. Contraction of the **arrector pilli muscle** causes the hair to rise to a more vertical position and increases the layer of insulating air against the skin. When you get goosebumps, it's the arrector pilli at work.

Glandular skin

The various glands in mammalian skin are one of two types: **sebaceous glands** or **sweat glands**. Hair is composed of dead cells, and if it were not "conditioned" in some way it would become brittle, break, or loose its waterproofing properties. To prevent this, sebaceous glands secrete oils and waxes, **sebum,** into the follicle, and the hair acts like a wick drawing up the secreted sebum, hairs' "natural" conditioner.

As their name implies, sweat glands produce watery sweat squeezed to the surface by the myoepithelium, which lines the ducts of the gland. Sweat glands are usually restricted to specific locations on a mammal's body and that location depends on the species. **Eccrine sweat glands** produce a watery fluid secreted onto the surface of the skin, and as it evaporates, it cools the body. **Apocrine sweat glands** look the same as eccrine glands but they release a thicker and often scented fluid into the hair follicle instead of the skin surface. The scents and smells produced by apocrine glands are important for communications in mammals.

The last type of gland gives the class Mammalia their name, **mammary glands**. The structure of the mammary gland resembles sebaceous glands but they have a myoepethelium like sweat glands. Whatever their origin, the glands secrete a nutritive fluid of fats, carbohydrates, and proteins used to feed the young. In the most primitive mammals, the monotremes, the mammary glands are grouped in a milk patch on the surface of the skin. To feed, a young monotreme licks the skin surface of the parent. Both male and female monotremes have functional mammary glands! Teats or nipples can be found in all other mammals and they form a long milk ridge that runs down the ventral side of the developing fetus. How many depends on the animal and the number of young that it produces.

Heterodent diphyodont dentition

In early vertebrates teeth were uniform cone-shaped structures embedded in the jaw. Mammals have a different type of dentition and specialized teeth with different functions. The presence of **incisors**, **canines**, **premolars,** and **molars** usually matches the food and feeding strategy of the mammal.

✍ *For what is each type of tooth modified?*

Mammals also have not one, but two sets of teeth during their lives. This is the **diphyodont** condition. The first are the milk teeth, and they will be replaced by the permanent teeth. As the animal grows, the jaw enlarges. When the second set of teeth are added, the jaw has finished growing. Molars only appear with the permanent teeth.

Muscular diaphragm

In mammals the floor of the thoracic cavity is lined with a muscular **diaphragm,** which, at rest, is dome-shaped. When the muscles in the diaphragm contract, it flattens, increasing the size of the thoracic cavity. At the same time, the **intercostal muscles** raise the rib cage, increasing the size of the thoracic cavity. The lungs are suspended in the cavity and the negative pressure created by the expanding thoracic cavity pulls air into them. Once inside the lungs, gas exchange occurs in the **alveoli,** which increase the available surface area for gas exchange. The **dual circuit circulatory system** is complete in mammals, and the pulmonary and systemic circuits are completely separated in the heart.

Extended care of young

The reproductive strategies of animals can be represented by two extremes. One is to produce huge numbers of eggs and hope that a few of them will survive to be reproductive adults. The opposite strategy is to produce only a few young and help them on their way to adult maturity. The first is the **r-selection** strategy and the second is the **K-selection** strategy. The thousands of eggs that a female insect lays is a good example of the r-selection strategy. Mammals use the K-selection strategy. Mammalian females produce only a few young during each reproductive cycle. Their young are protected internally as they grow and after their birth stay with the parent and are fed using the mammary glands. During **suckling**, and **weaning** that follows, young mammals are protected and learn a variety of behaviors and habits important for their survival. The lengths of time that the young are cared for, and how mature they are when born, varies in different mammals.

What may surprise you is that not all mammals are **viviparous**. The most primitive are the **oviparous** monotremes that lay eggs. Marsupials go a step farther, and the young embryo develops in the uterus for a short time before crawling out and into a protective pouch where the mammary glands are found. The last group of mammals are the placental mammals, and the developing embryo uses the **placenta** to exchange nutrients, oxygen, and metabolic wastes with the mother.

A Closer Look Inside *Digital Zoology* - Mammalia

The fetal pig

Mammals, like birds, have completely separated the pulmonary and systemic circuits of the circulatory system. Oxygen-poor blood that has circulated through the body enters the right atrium of the heart through the **caudal vena cava** and **cranial vena cava** from the lower and upper parts of the body, respectively. From there it goes to the right ventricle, which pumps it into the **pulmonary artery** and on to the lungs for oxygenation. The oxygen-rich blood returns to the heart's left atrium through the **pulmonary vein**, passes to the left ventricle, and is then pumped out to the body through the **aortic arch**. The first major branch to come off the aortic arch is the **brachiocephalic artery,** which forms the **right subclavian artery** and the **common carotid artery,** which will branch to form the left and right common carotids that branch even farther as they supply the head. The second branch to come off the aortic arch is the **left subclavian artery** completing the pair of arteries that supply the anterior part of the body and the forelimbs. The aortic arch forms the **dorsal aorta** that runs the length of the animal, and its main branches supply different organs and parts of the body: the stomach, pancreas and spleen, with the **celiac artery**; the liver, the **hepatic artery**; the mesenteries, the cranial and caudal **mesenteric arteries**; the adrenal gland, the **adrenal arteries**; the kidneys, the **renal arteries**. **Genital arteries** branch off the dorsal vessel to supply gonads, and the posterior end of the dorsal aorta branches to form the **common iliac artery,** which supplies blood to the hindlimbs.

In most cases, blood from the lower part of the body returns to the heart through a set of veins corresponding to the arteries that supplied the region. It's a little more complicated in the anterior region where external and internal **jugular veins** from the head join with the **cepahlic veins** and **subclavian veins** on each side to form the left and right **branchiocephalic veins**. These then fuse to form the cranial vena cava, which joins the caudal vena cava at the right atrium of the heart. The digestive system doesn't return its blood directly to the caudal vena cava like other organs do. Instead, blood from the alimentary tract passes to the liver through the **hepatic portal system**. Only after it has passed through the capillary beds of the liver does it return to the caudal vena cava on its way back to the heart.

✍ *What is the role of the hepatic portal system?*

At some point during embryonic development each of the original six aortic arches makes their appearance, three will disappear during development, and only three will remain when the mammal is born: the **third aortic arches** forming the internal carotid arteries. The large aortic arch is formed from the left side of the **fourth aortic arch** and the right hand side forms the subclavian artery. Ancestrally both of these aortic arches would have connected with the dorsal aorta but in mammals only the left side makes the connection. The **sixth aortic arches** supply the lungs.

A fetus' lungs aren't functional until after it's born, and there's no point sending large volumes of blood through the pulmonary circuit prior to that. The **ductus ateriosus,** which connects the pulmonary artery to the dorsal aorta, short circuits the system. Blood that would otherwise have gone to the lungs goes to the dorsal aorta and into the systemic circuit. This is the first of two short circuits in the fetal circulatory system. The second is inside the heart. The **foramen ovale**, an opening between the right and left atriums, allows blood returning from the body that enters the right atrium, to flow into the left atrium and from there to the left ventricle and back out to the body.

A fetus depends entirely on the placental blood supply for oxygen. An umbilical vein from the placenta connects to the fetal circulatory system through the **ductus venosus,** which passes blood to the liver and the caudal vena cava. From there it goes directly to the fetal heart. Blood returns to the placenta through the umbilical arteries connected to the posterior end of the dorsal aorta. When the young mammal is born, placental circulation stops, and with its first breath, a muscle inside the ductus arteriosis closes that vessel and fibrous tissue seals the tube forming a ligament that connects the dorsal aorta and pulmonary artery. Muscles also contract and close the umbilical arteries, which form ligaments or the common and internal iliac arteries.

Cross-References to Other McGraw-Hill Zoology Titles

Integrated Principles of Zoology, 11th edition. C.P. Hickman, L.S. Roberts & A. Larson. Chapter 30.

Animal Diversity, 2nd edition. C.P. Hickman, L.S. Roberts & A. Larson. Chapter 19.

Zoology, 5th edition. S.A. Miller & J.P. Harley. Chapter 22.

Vertebrate Biology, 1st edition. D.W. Linzey. Chapter 9.

Laboratory Studies in Integrated Principles of Zoology, 10th edition. C.P. Hickman, F. Hickman & L. Kats. Chapter 22.

General Zoology Laboratory Guide, 13th edition. C. Lytle. Chapter 19 & 20.

Structures Checklist

Here are some of the structures that you should be able to easily find in *Digital Zoology* and the specimens that you will be looking at in your lab. After reading your lab handout, you might want to add more and, depending on the equipment available in your lab, you might see more. As you study the material, you might also want to make some notes on how some of these structures looked or include a drawing in your lab notes. (Structures indicated by * may be hard to see.)

Fetal Pig - External anatomy and muscular systems

External anatomy

- ☐ Ankle
- ☐ Anus
- ☐ Auricle (Ear)
- ☐ Digits
- ☐ Elbow
- ☐ External nares (Nostrils)
- ☐ Eyelids
- ☐ Eyes
- ☐ Genital papilla
- ☐ Head
- ☐ Hooves
- ☐ Knee
- ☐ Mammary papilla
- ☐ Neck
- ☐ Nictitating membrane
- ☐ Scrotal sacs
- ☐ Shoulder
- ☐ Tongue
- ☐ Trunk
- ☐ Umbilical cord
- ☐ Urogenital opening
- ☐ Vibrissae
- ☐ Wrist
- ☐
- ☐
- ☐
- ☐
- ☐

Musculature - Head and Trunk

- ☐ Brachiocephalicus
- ☐ Digastric
- ☐ External oblique
- ☐ Internal oblique
- ☐ Linea alba
- ☐ Masseter
- ☐ Mylohyoid
- ☐ Omohyoid
- ☐ Rectus abdominis
- ☐ Rectus thoracis
- ☐ Sternohyoid
- ☐ Sternomastoid
- ☐ Sternothyroid
- ☐ Temporalis
- ☐ Thyrohyoid
- ☐
- ☐
- ☐
- ☐

Musculature - Pectoral and forelimb

- ☐ Biceps
- ☐ Brachialis
- ☐ Cleidomastoid
- ☐ Cleidooccipitalis
- ☐ Coracobrachialis
- ☐ Deltoideus
- ☐ Extensor carpi radialis
- ☐ Extensor digitorum communis
- ☐ Extensor digitorum lateralis
- ☐ Flexor carpi radialis
- ☐ Flexor carpi ulnaris
- ☐ Flexor digitorum superficialis
- ☐ Flexor digitorum profundus
- ☐ Latissimus dorsi
- ☐ Omotransversarius
- ☐ Pectoralis profundus
- ☐ Pectoralis superficialis
- ☐ Rhomboideus capitis
- ☐ Rhomboideus cervicis
- ☐ Rhomboideus thoracis
- ☐ Serratus ventralis
- ☐ Serratus dorsalis
- ☐ Supraspinatus
- ☐ Teres major
- ☐ Trapezius
- ☐ Triceps
- ☐ Ulnaris lateralis
- ☐
- ☐
- ☐

Musculature - Pelvic and hindlimb

- ☐ Adductor
- ☐ Biceps femoris
- ☐ Extensor digitorum longus
- ☐ Flexor digitorum longus
- ☐ Gastrocnemius and soleus
- ☐ Gluteus medius
- ☐ Gluteus profundus
- ☐ Gluteus superficialis
- ☐ Gracilis
- ☐ Iliacus
- ☐ Pectineus
- ☐ Peroneus longus
- ☐ Peroneus tertius
- ☐ Psoas major
- ☐ Quadratus femoris
- ☐ Rectus femoris
- ☐ Sartorius
- ☐ Semimembranosus
- ☐ Semitendinosus
- ☐ Tensor fasciae latae
- ☐ Tibialis anterior
- ☐ Tibialis posterior
- ☐ Vastus lateralis
- ☐ Vastus medialis
- ☐
- ☐
- ☐
- ☐

Additional structures

- ☐
- ☐
- ☐
- ☐
- ☐

Fetal Pig - Internal anatomy

Digestive and related systems

- ☐ Anus
- ☐ Cecum
- ☐ Common bile duct
- ☐ Cystic duct
- ☐ Duodenum
- ☐ Epiglottis
- ☐ Esophagus
- ☐ Gall bladder
- ☐ Glottis
- ☐ Hard palate
- ☐ Ileum
- ☐ Lesser omentum
- ☐ Liver
- ☐ Mandibular gland
- ☐ Mesentery
- ☐ Nasopharynx
- ☐ Pancreas, left lobe
- ☐ Pancreas, right lobe
- ☐ Parotoid duct
- ☐ Parotoid gland
- ☐ Rectum
- ☐ Rugae of stomach
- ☐ Jejunum
- ☐ Soft palate
- ☐ Spiral colon
- ☐ Spleen
- ☐ Stomach
- ☐ Sublingual gland
- ☐ Teeth
- ☐ Thymus
- ☐ Thyroid gland
- ☐ Tongue
- ☐
- ☐
- ☐
- ☐
- ☐

Circulatory and respiratory

- ☐ Aortic arch
- ☐ Axillary vein
- ☐ Axillary artery
- ☐ Brachiocephalic trunk
- ☐ Brachiocephalic vein
- ☐ Caudal mesenteric artery
- ☐ Caudal vena cava
- ☐ Celiac artery
- ☐ Cephalic vein, left and right
- ☐ Common carotid artery
- ☐ Coronary artery
- ☐ Coronary vein
- ☐ Costocervical trunk
- ☐ Cranial mesenteric artery
- ☐ Cranial vena cava
- ☐ Deep circumflex artery
- ☐ Deep circumflex vein
- ☐ Deep circumflex iliac artery
- ☐ Deep circumflex iliac vein
- ☐ Deep femoral artery
- ☐ Deep femoral vein
- ☐ Diaphragm
- ☐ Dorsal aorta
- ☐ Ductus arteriosus
- ☐ External iliac artery
- ☐ External iliac vein
- ☐ External jugular veins
- ☐ Femoral artery
- ☐ Femoral vein
- ☐ Gastroduodenal artey
- ☐ Genital artery
- ☐ Genital vein
- ☐ Heart
- ☐ Hepatic artery
- ☐ Internal iliac artery
- ☐ Internal iliac vein
- ☐ Internal jugular veins
- ☐ Larynx
- ☐ Left atrium
- ☐ Left ventricle
- ☐ Lungs
- ☐ Median sacral artery
- ☐ Median scaral vein
- ☐ Mesenteric arteries
- ☐ Mesenteric veins
- ☐ Portal vein
- ☐ Primary bronchi
- ☐ Pulmonary arteries
- ☐ Pulmonary trunk
- ☐ Pulmonary veins
- ☐ Renal artery
- ☐ Renal vein
- ☐ Right atrium
- ☐ Right ventricle
- ☐ Splenic artery
- ☐ Splenic vein
- ☐ Subclavian arteries
- ☐ Subclavian veins
- ☐ Subscapular vein
- ☐ Trachea
- ☐ Umbilical arteries
- ☐ Umbilical vein
- ☐
- ☐
- ☐
- ☐
- ☐

Urogenital systems

- ☐ Bulbourethral glands
- ☐ Cremaster pouch
- ☐ Ductus deferens
- ☐ Epididymis
- ☐ Genital papilla
- ☐ Gubernaculum
- ☐ Horns of the uterus
- ☐ Kidney
- ☐ Medula
- ☐ Ovary
- ☐ Oviduct
- ☐ Penis
- ☐ Prostrate
- ☐ Renal calyces
- ☐ Renal pelvis
- ☐ Renal pyramid
- ☐ Round ligament
- ☐ Scrotum
- ☐ Spermatic cord
- ☐ Suprarenal gland
- ☐ Testis
- ☐ Ureter
- ☐ Urethra
- ☐ Urinary bladder
- ☐ Vagina
- ☐
- ☐
- ☐
- ☐
- ☐

Additional structures

- ☐
- ☐
- ☐
- ☐
- ☐
- ☐
- ☐

Crossword Puzzle - Mammalia

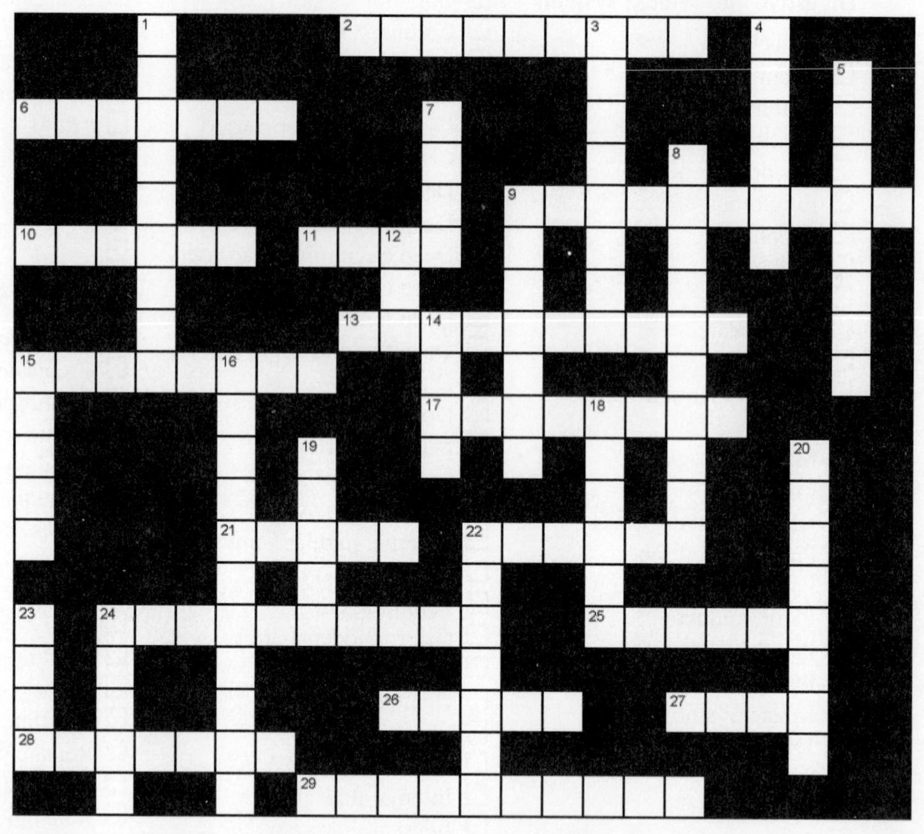

Puzzle solution. An interactive web based version of this puzzle, and its solution, are available on the *Digital Zoology* web site at www.mhhe.com/DigitalZoology/Students. With the interactive puzzle you can check to see if individual words or the whole puzzle is correct, and get hints for single letters.

Across

2 Describes the soft, dense, insulative hairs. (5,4)

6 These teeth, which you would find in predatory mammals, are missing in an herbivore. (7)

9 This type of mammal carries its young in a pouch. (10)

10 In placental mammals the eggs develop in this part of the female reproductive tract. (6)

11 Flying mammals. (4)

13 These are egg-laying mammals. (10)

15 What we're more likely to call the vibrissae, the sensory tactile hairs in mammals. (8)

17 Of the three skull types in vertebrate, mammals have this one. (8)

21 Mammalian taxonomists spend a lot of time looking at mammal jaws and these when dividing the class into different taxa. (5)

22 These teeth are important for crushing and grinding food. (6)

24 Birds and mammals have to maintain their body temperature at a fixed point with little or no variation. (10)

25 In mammals the cycle of female fertility. (6)

26 These hairs in mammals protect the fur and provide its coloration. (5)

27 The number of different types of glands that you'll find in mammalian integument. (4)

28 Major protein of mammalian hair. (7)

29 At some point in every mammal's life the first set of teeth fall out and are replaced with a second set. This is the term for this type of dentition. (10)

Down

1 Another term for egg-laying animals. (9)

3 The skin glands that develop at the onset of sexual maturity in mammals. (8)

4 Compared to other vertebrates, this part of the integument is larger and is the biggest part of the mammalian integument. (6)

5 Hairs in mammals grows out of this. (8)

7 These secretions of the sebaceous gland help maintain the hair of mammals. (4)

8 The viviparous placental mammals. (10)

9 Many mammals replace their coat of hair twice a year in a process referred to as shedding or this. (7)

12 The number of sets of teeth that diphyodont mammals have. (3)

14 The only place will you'll find the ancestral mammalian hairs in a whale. (4)

15 What an eccrine gland in a mammal secretes. (5)

16 Metabolism warms the bodies of mammals and birds. This term describes that type of heat. (11)

18 Another term for a mammal's first coat. (6)

19 These mammalian glands are used for communication; they're especially large in skunks so that you can be sure to get the message. (5)

20 The sharp edges of these mammalian teeth make them ideal for cutting and biting. (8)

22 Modifications of the apocrine glands to nourish the young are one of the possible origins of these glands in mammals. (7)

23 A mammals first set of teeth are called deciduous, or this type of teeth. (4)

24 Like feathers, these help insulate mammals and are a diagnostic character of this vertebrate class. (5)

Self Test - Mammalia

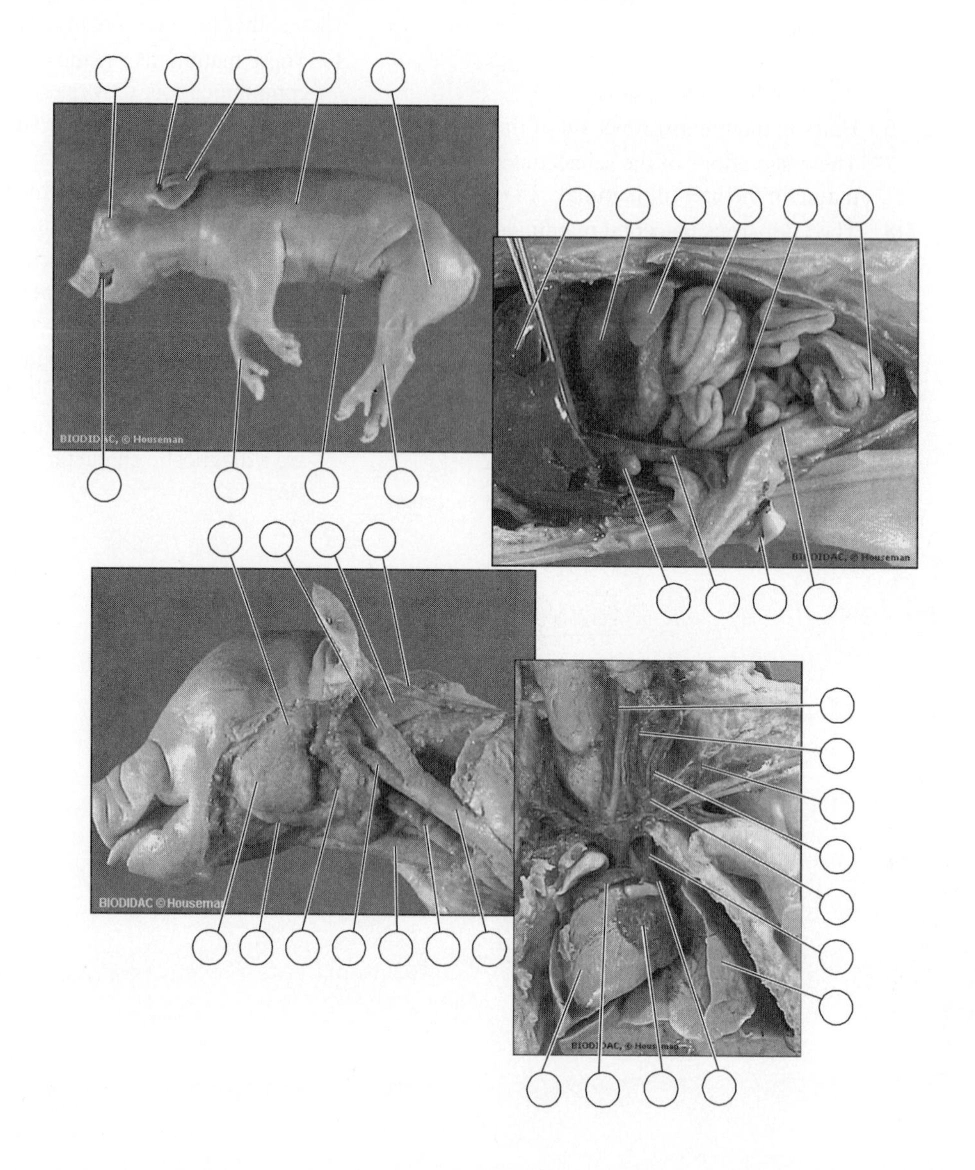

Use the following labels to identify the photographs. You may have to use a label more than once, and some labels may not be appropriate for the photographs. Answers are available on the *Digital Zoology* web site at www.mhhe.com/digitalzoology. Be sure to try the interactive Drag-and-Drop quizzes that are available on the *Digital Zoology* CD-ROM. A color version of this Self Test is available in the Adobe Acrobat version of the Student Workbook in the workbook folder on the *Digital Zoology* CD-ROM.

Specimen

Fetal Pig

Labels

1) Ankle
2) Aortic arch
3) Auditory canal
4) Brachiocephalic vein
5) Brachiocephalicus
6) Cephalic vein
7) Cleidomastoid
8) Cleidooccipitalis
9) Common carotid artery
10) Ear
11) External jugular vein
12) Eye
13) Gall bladder
14) Hindlimb
15) Ileum

16) Internal jugular vein
17) Left auricle
18) Left ventricle
19) Liver
20) Masseter
21) Parotid duct
22) Parotid gland
23) Rhomboideus capitis
24) Right auricle
25) Spiral colon
26) Spleen
27) Splenius
28) Sternohyoid
29) Sternomastoid
30) Stomach
31) Subclavian artery
32) Temporalis
33) Thymus gland
34) Thyroid gland
35) Tongue
36) Trunk
37) Umbilical artery
38) Umbilical cord
39) Umbilical vein
40) Wrist

Glossary

4d Cell In animals with determinate cell cleavage, this cell, which appears at the 64 cell stage, forms all of the mesodermal structures.

Abdomen In organisms that have undergone tagmosis the trunk often develops into two tagma, one involved in locomotion and one that is not locomotory. The tagma not involved in locomotion is referred to as the abdomen.

Aboral surface In radially symmetric animals there is no anterior, posterior, or right sides to the animal. Instead we refer to the two sides of the animal by referring to the location of the mouth. In this case, the side opposite the mouth.

Acanthus The egg and larval stage of an acanthocephalan that passes from the female parasite to the feces of the host.

Acetabulum One of two suckers found on the digenic flukes (trematodes). The oral sucker surrounds the mouth, and the acetabulum is located on the ventral surface.

Acoelomate Triploblastic animals that do not have an internal body cavity. This includes the flatworms and ribbon worms. Although the term could be applied to other lower phyla, it is most accurately used with the triploblasts rather than diploblasts.

Adaptive radiation Evolution of a variety of different species from a single common ancestor. Each is adapted for a particular niche, and the appearance of the descendants may vary considerably from each other and the ancestor.

Advanced characters The traits, or characteristics, that an animal has that are not ancestral to the taxon. They appear later in the evolutionary history of the taxon and cannot be found on the ancestor to the group.

Akinetic skull The type of skull where there is no joint that allows the upper jaw and palate that form the snout or nose to move relative to other parts of the skull.

Allantois One of the extraembryonic eggs found in the amniote animals. The allantois contains the metabolic wastes created by the developing embryo. May also be involved in gas exchange and is important in forming the placenta in placental mammals.

Ambulacral groove The groove that runs down the oral surface of each echinoderm arm and contain the tube feet. If the region contains a visible furrow, or groove, they are referred to as open, if not, then closed.

Amnion One of the extraembryonic membranes found in terrestrial vertebrate. The amnion is filled with fluid and the developing embryo of reptiles, birds, and mammals develops suspended inside the amnion.

Amniote egg Eggs that shelter the developing embryo in a water-filled sac—the amnion. Characteristic of the amniote animals, which include reptiles, birds, and mammals.

Amphid Unique to the nematodes, these paired sensory structures are located on the sides of the head. An external pore leads to an inner chamber and the sensory receptor formed from modified cilia.

Analogy, Analogous Refers to structures that do not have the same evolutionary origin but have the same functions. The wings of a bat and insect are analogous because both are used for flight. The two structures, however, are not formed from a common structure.

Anapsid The type of skull found in reptiles and seen today only in turtles. Behind the opening of the orbits the skull is solid and lacks temporal openings found in other amniote vertebrate skulls.

Ancestral characteristic A character shared by all members of a taxonomic group of organisms (taxon) and used to define the unique nature of the group. The character may be modified or even disappear in some members of the taxon.

Antennae Sensory appendages found on the head of a variety of animals, although we commonly associate them with uniramians and crustaceans.

Antennule The small sensory biramous appendages found on the first segment of crustaceans.

Apical complex Specialized organelles in parasitic protozoans in the phylum Apicomplexa believed to help the parasite penetrate and enter the host so that they can complete their life cycle.

Appendicular skeleton The bones of the arms, wings, legs, and fins of vertebrates, along with the pelvic and pectoral girdles when present, that attach the limbs to each other and then in turn to the axial skeleton.

Aquiferous system This type of system is found in sponges and is consists of the canals and chambers through which water flows. Water is pumped through the system by the choanocytes.

Archenteron The name given to the primitive gut, the first tube that runs through the developing embryo and is open to the external environment. Formed during gastrulation, it is surrounded by the new endoderm and it will develop into the digestive system of the organism.

Aristotle's lantern The specialized feeding structure is found in the echinoid echinoderms, sea urchins, and sand dollars. Consisting of 40 different ossicles the five teeth are used in feeding.

Arthropodization Many of the traits that we consider unique to the Arthropoda may not be independent traits. Instead, they may be the consequence of a tubular exoskeleton. If this is the case, the traits result from a single event, arthropodization—the acquisition of an exoskeleton.

Asconoid sponge Of the three different sponge architectures, this is the simplest. It consists of a central choanocyte lined spongocoel that opens to the outside directly through the osculum. Water enters the spongocoel after passing through the dermal pores.

Asexual Organisms that do not reproduce by recombination of genetic material contained in gametes. There is no combination of a haploid sperm and egg to form a zygote.

Asymmetric body plan A body plan where there is no axis of symmetry that runs through the body and creates identical parts.

Atriopore The external opening to the atrium. In cephalochordates water passes across the pharyngeal slits into the atrium and from there leaves through the atriopore.

Auricle Chamber of the heart that receives the blood from outside the heart.

Axial skeleton The bones that make up the skeleton of the main body axis of vertebrates. It includes the cranium, vertebral column, and the rib cages, although not all of these may be present in each of the vertebrate groups.

Bilateral symmetry In organisms that have a bilaterally symmetric organization there is only one way that the axis of symmetry can pass through the oral aboral axis and create two identical halves.

Binary fission Cell division where the parent cell divides into two daughter cells equal in size to each other.

Biogenetic Law In short, ontogeny recapitulates phylogeny. Or, the embryological or developmental sequence of an animal repeats the evolutionary history that gave rise to that animal.

Bipedal locomotion Walking upright using the pelvic girdle and limbs.

Biradial symmetry The organism appears radially symmetric but at least one set of structures is paired. This results in only two planes of symmetry that pass through the oral-aboral axis of the animal.

Biramous A type of appendage in arthropods. Biramous appendages are branched and Y-shaped.

Blastocoel During the development of the embryo, single cells divide and form a hollow ball of cells one cell layer thick. The cavity inside this hollow ball is the blastocoel.

Blastopore The opening to the primitive gut (archenteron) that will develop into either the mouth or anus. The blastopore forms during gastrulation.

Bony fish The common name for fish in the vertebrate class Osteichthyes. Unlike the cartilaginous fish the skeleton is hardened by the deposition of calcium salts.

Book gill Respiratory structures found in some aquatic cheliciformes. They consist of multiple sheets, or lamellae, across which water flows.

Book lung The respiratory structures found in some spiders. It consists of multiple folds of the cuticle that form sheets, or lamellae, across which air moves for gas exchange.

Branchial basket In ascidian adults the pharynx is enlarged to form the branchial basket. Pharyngeal openings in the wall of the basket allow water to pass through with food being trapped on the inner surface before being passed into the gut.

Branchial heart Modification of the circulatory system to form secondary hearts that pump blood into the gills.

Budding A form of asexual reproduction where a small part of the body separates from the parent and develops into a complete organism.

Bursae Paired water-filled sac located at the base of each arm of a brittle star. Cilia in the bursae circulate water. Its surface is important in gas in gas exchange. Gametes may also be released into the bursa.

Captacula The specialized ciliated tentacles that scaphopod molluscs use for feeding. Food is trapped on the sticky bulb at the tip of the captaculum before being moved to the mouth by cilia on the captacula.

Carapace The protective dorsal shield or plate that covers an animal. For example in the Crustacea the cuticle fuses to form a plate over the cephalothorax. In turtles the whole shell is referred to as a carapace.

Caudal A tail or the tail region.

Caudal rami In primitive crustacea the last segment of the body often has paired limblike branches on it. They may be flexable or solid and spinelike.

Cellular grade Organisms with this type of cellular organization are referred to as the parazoan. They have distinct cells that function independently of each other even though some cells may take on specialized functions. Groups of cells never work together and function as a tissue.

Cephalization Evolution of a distinct anterior region of the body, the head, with specialized sensory structures.

Cephalothorax In many arthropods the appendages found on the thorax become involved in feeding, and when this happens they are referred to as maxillipeds. The new tagma that results is the cephalothorax.

Cercaria A stage in the life cycle of trematode flukes. The cercaria develops from redia found in the intermediate host. This tadpolelike organism is released from the intermediate host to locate either the primary host or another intermediate host."

Chelicera The unique paired feeding appendage on the first segment of the body of Cheliciformes such as spiders and ticks.

Chelicifore These feeding mouthparts are found on most, pycnogonads. Although the proboscis is the main feeding structure in these animals when chelicifores are present, they either hold or tear food and pass it to the proboscis.

Chevron Having a V-shape.

Chitin A complex carbohydrate composed of linearly arranged N-acetyl-glucosamine units. Chitin is a characteristic of the cell wall of fungi and the outer cuticle of arthropods.

Chitinous A structure composed of chitin, a complex carbohydrate composed of linearly arranged N-acetyl-glucosamine units. Chitin is a characteristic of the cell wall of fungi and the outer cuticle of arthropods

Choanocyte The unique collar-shaped cells whose flagella are responsible for generating the water current in the sponge. As the flagella beats, food particles are trapped against the microvilli that form the collar. Choanocytes are also found in some colonial protists.

Choanoflagellate The protozoans that resemble the choanocyte cells found in sponges. They have a collar of microvilli that surround a central flagellum. When the flagellum beats, food is trapped against the collar and consumed by phagocytosis.

Chorion The outermost extraembryonic membranes in reptiles, birds, and mammals. It is involved in gas exchange. In insects, the chorion is the outer shell of the egg secreted by the follicle cells of the ovary.

Cilia A cellular hairlike locomotory structure that consists of an extension of the plasma membrane surrounding a 9+2 organization of microtubules. Unlike flagella, cilia are shorter and more numerous on the cell surface.

Clade In cladistic taxonomy, how groups of animals are related to each other. It includes the ancestral group and all of the other groups, or lineages, of animals that arose from the ancestral group. Clades are monophyletic.

Cladistics A method for classifying organisms based on primitive and derived characters. The resulting arrangement of organisms reflect evolutionary relationships between the taxa.

Cladogram A visual representation of the phylogenetic branching of different animal groups based on cladistics. There is no units or measures to the dimensions of a cladogram.

Clitellum Special fused metameres in oligocheates and leeches that secrete mucous during mating.

Cloaca The shared common opening to a number of organ systems that may include the reproductive, digestive, and/or excretory system.

Cloacal siphon Part of the ascidian body that surrounds the point where the water leaves the animal—excurrent flow. Water is pumped the action of the cilia lining the pharyngeal basket.

Closed circulatory system A circulatory system in which the circulating fluid (blood) flows in vessels or tubes connected to each other by capillaries.

Cnidocyte Specialized cells found only in the Cnidaria. When these cells evert, a nematocyst is discharged. The nematocyst may act as a stinger or a sticky thread to entangle and capture prey.

Coelenteron The name given to the internal cavity of the cnidarians. This is an incomplete gut with only one opening, the mouth. Food to be digested and undigested food that must be eliminated both pass through the mouth. The cavity is lined by gastrodermis.

Coelom A true body cavity completely lined by mesoderm, which forms the peritoneum. Animals with true coeloms are referred to as eucolomates.

Coelomoduct Tubules that connect the coelomic cavity with the outside of the animal.

Collagen Tough fibrous protein found in the connective tissue of vertebrates and the cuticles of some invertebrates. The protein is flexible but doesn't stretch or compress.

Colloblast cell The unique sticky discharge cells found in the tentacles of ctenophorans. They are used to capture prey, and although cnidoblasts resemble cnidocytes, this is a convergent trait making the two analogous structures.

Collum Unique structure in millipedes. It is formed from the first trunk segment, which has lost its appendages and become enlarged into a collarlike segment that surrounds the head.

Complete alimentary tract, Complete digestive tract, Complete gut An digestive tract that has an anal and oral opening. This adaptation allowed for a linear processing of ingested food with specialized regions in the gut for grinding, mixing, and digesting food under different conditions.

Complex ciliature Specialized ciliary structures found in the higher ciliates (Ciliophora). The single 9+2 cilia fuse with others to form larger more complex structures such as cirri and membranelles.

Compound eye The characteristic eye of the Arthropoda. It consists of many ommatidia, the basic optical unit of the compund eye, grouped together to form the compound eye. Compound eyes are usually found in combination with much simpler single lens ocelli.

Conjugation A form of sexual reproduction in ciliates (Ciliophora). During conjugation two ciliate protozoans join and the macronuclei disappears. After meiotic divisions of the micronucleus, the resulting genetic material is exchanged.

Convergent evolution When two distinct evolutionary lineages do not share a common ancestor but evolve similar structures or features. These features are considered analogous.

Convergent trait A trait or characteristic similar in two animals from two individual evolutionary lineages. These characters are considered analogous.

Corona The crown formed from a ciliated disc at the anterior end of a rotifera.

Coxal gland The excretory gland found in some spiders. This blind-ended sac is bathed in hemolymph, and wastes are excreted at the base of the coxal segment of the leg.

Cranium The brain case, made of either cartilage or bone, that surrounds the brain of vertebrates.

Cryptobiosis A dormant, or suspended, state of animation that allows some animals to survive severe conditions such as extremely low temperatures. Dehydration occurs in preparation for the suspended state and rehydration is associated with the return of favorable conditions.

Crystalline style A rodlike structure in some mollusc stomachs made of enzymatic proteins required for digestion. Cilia lining the stomach rotate the crystalline style, which grinds against the gastric shield to release the digestive enzymes.

Ctene This unique locomotory structure is found in the Ctenophora and is made up of fused cilia arranged into flattened plates. The ctenes are then organized into eight bands that run between the oral to aboral surface of the animal.

Ctenidia Molluscan gills that often have additional functions other than respiration.

Cuticle The nonliving, and noncellular outer layer of an organism secreted by the underlying epidermis. Cuticles are common in a variety of animals including nematodes, annelids, and arthropods. The presence of a cuticle precludes the presence of cilia.

Cuvierian tubules Defensive organs used by some sea cucumbers. They are located at the base of the respiratory tree and are shot out of the cloaca if the animal is threatened.

Cytostome In some protozoans, especially ciliates, phagocytosis always occurs at the same position on the cell surface. When this occurs the location is referred to as the cytostome (cell mouth).

Dendogram A visual representation of evolution that branches like a tree to depict the relationships between the different groups.

Dermal branchiae External extensions of the outer epidermis and peritoneum of the echinoderm body cavity. Both the outer epidermis and inner peritoneum are lined with cilia. The surface of the dermal branchiae are important in diffusion of gases and metabolic wastes.

Deuterostome Phyla, including the Chordata and Echinodermata, that share common characteristics of the blastopore—not forming the mouth, radial indeterminate cellular cleavage in the embryo, and the formation of the body cavity by enterocoelic pouching.

Diapsid Vertebrate skull of some reptiles and all birds. There are two temporal openings behind the orbits in the skull.

Digestive caecum A blind-ended pouch that extends from the main digestive tract. Digestive ceca may be the sites for final digestion of ingested food or may be regions of the gut with specialized enzymes or conditions required for digestion.

Digestive gland Many invertebrates, including molluscs, arthropods, and echinoderms, have pockets or evaginations from the main alimentary tract where specialized digestive events occur.

Dimorphic life cycle, Dimorphism When the life cycle of the animal includes two distinct and physically different body types. The life cycle progresses from the immature form to the reproductive type. Or both forms may appear together in colonial types of dimorphic organisms.

Dimorphic nuclei Organization of the nuclear material typical of ciliates. There is always a single small micronucleus and either one large or many macronuclei. Micronuclei are involved in reproduction and the macronuclei are used by the cell to function.

Dioecious Organisms that have the male or female reproductive structures in separate individuals. This is the opposite of monoecious.

Diphyodont teeth The type of teeth found in mammalian vertebrates with two sets of teeth—a baby set replaced with permanent adult teeth.

Diploblastic Organisms formed from only the two primitive cell layers—endoderm and ectoderm. Although there may be some type of a matrix between the two cell layers, often referred to as mesoglea or mesenchyme, it is not a true tissue layer.

Diplosegments A characteristic of the millipedes, diplopoda, where adjacent segments have fused in pairs. As a result there appears to be two pairs of appendages for each of the visible segments of the trunk.

Diverticulum Any hollow branch or pouch that extends from the side of an organ.

Dual circulatory system In vertebrates such as amphibians, reptiles, birds, and mammals, blood flows through two different paths or circuits. One to the lungs, pulmonary, and the other to the rest of the body, systemic.

Ecdysis The periodic molting, or shedding, of the outer exoskeleton of an arthropod.

Ectoderm The outermost cell layer that forms the epithelium and nervous systems of an animal. It, and the endoderm, are the two primary germ layers.

Ectoparasite, Ectoparasitism Parasitic organisms depend on another animal, their host, for survival. Ectoparasites live on the outer surface of the host. This is the opposite of an endoparasite, which lives inside the host.

Ectotherm Organisms unable to generate body heat internally using metabolic processes and instead use their environment. The opposite is endothermic.

Elastic capsule chromatophore Specialized pigmented cells on the surface of cephalopods that, by changing their shape, expose differing amounts of pigment changing the color and appearance of the cephalopod.

Endoderm The innermost layer of cells that forms the digestive tract and other associated organs. The ectoderm and the endoderm form the two primary germ layers of an animal.

Endoparasite Parasitic organisms that depend on their host for all, or at least one stage, of their life cycle. Endoparasites complete this requirement inside the host.

Endopodite The inner branch of the biramous appendage in crustaceans.

Endoskeleton Supporting structures of the skeleton surrounded by the body tissues. As a consequence, there is living tissue on all sides of endoskeletal structures. Endoskeltons are found, for example, in echinoderms and chordates.

Endostyle A ciliated groove on the ventral surface of the pharynx in early chordates and related taxa. Mucous produced by the endostyle traps particulate food, and the cilia propel it into the digestive tract.

Endotherm, Endothermic Organisms able to generate body heat internally using metabolic processes. The opposite is ectotherm.

Epicuticle The outermost part of the arthropod cuticle that lacks any chitin. The epicuticle forms a chemical barrier, and the waxes in it prevent cuticular water loss in terrestrial arthropods.

Epistome Flaplike covering of the oral opening in lophophores. Formed from the protocoel, it may or may not have a coleomic cavity inside.

Epitheliomuscular cell Cells that line the outer surface of cnidarians. These cell have two functions: the first is to form the outer body covering of the animal, the second is in movement by contraction of the myoneme portion of the cell.

Eucoelomate Animals that have a true coelom where the entire body cavity is lined with mesoderm.

Eutely Also referred to as cellular constancy. Eutelic animals are often composed of a fixed number of cells, or nuclei, in the adult, and this number is species specific. Eutely is often associated with animals that have specialized for miniaturization.

Evisceration A presumed defensive strategy used by sea cucumber. When threatened, they shoot parts of the digestive tract and other organ systems out the cloaca at their attacker.

Evolutionary taxonomy Based primarily on the principles of evolution. New taxa result from genetic isolation or niche specialization.

Exopodite The outer branch of the biramous appendage in crustaceans.

Exoskeleton Supporting structures of the skeleton not surrounded by the body. Endoskeletal elements are secreted by an underlying epidermis and one side of the skeletal structures is exposed outside the body. Found, for example, in arthropods.

Extant Species, or groups of animals, that still live today and have not become extinct.

Extinct Animals that existed in the past but are no longer living species.

Extracorporial Outside of the body. An example is extracorporial digestion found in some animals where digestive enzymes are regurgitated into the food. Once it has been broken down, the resulting product is consumed.

Extraembryonic membranes A set of membranes that surround the developing embryo in reptiles, birds, and mammals. They include the amnion, chorion, allantois, and yolk sac.

Flagella A cellular hairlike locomotory structure that consists of an extension of the plasma membrane surrounding a 9+2 organization of microtubules. Unlike cilia, flagella are longer and are usually found singly or in pairs on each cell.

Forelimb The appendages attached to the anterior pectoral girdle of vertebrates. The forelimbs may be modified as legs, arms, wings, or fins.

Gamete One of two types of sex cells, egg or sperm, that fuse to form the zygote.

Ganoid scale Thick, square fish scales that don't overlap. Found in primitive fishes.

Gastrodermis The name given to the endodermal cells that line the gastrovascular cavity (coelenteron) of cnidarians.

Gastrovascular cavity The name given to the internal cavity of the cnidarians. A blind-ended (incomplete) gut with only one opening. Food is to be digested and undigested food that must be eliminated pass through the mouth. The cavity is lined by gastrodermis.

Gastrulation During embryological development this stage results in the blastulas conversion into a gastrula. Cells migrate toward the inside of the embryo from the region where the blastopore will form to create the second germ layer (endoderm). The embryo changes from having only one cell layer to having two cell layers.

Gill arches Cartilagenous or bony supports for the gills in vertebrates. They form from the tissues between the pharyngeal slits.

Gill slits Lateral opening in the wall of the pharynx that allows water to enter into the mouth and exit through the pharynx. This is one of the ancestral characteristics of the phylum Chordata. It allowed water to be removed from ingested food before it was passed back into the digestive system.

Gnathobase In many arthropods the base of the appendages are modified into grinding surfaces used to break up food before it is passed to the mouth.

Gnathochilarium Structure formed by the fusion of the first maxillae. This flaplike structure is found in millipedes, diplopods, and pauropauds.

Gonad The reproductive organ that produces gametes, eggs, or sperm by meiotic reduction of the chromosome content.

Gravid An organism either filled with eggs or pregnant.

Hemerythrin A violet-colored respiratory pigment that uses iron in the oxygen binding site. Hemerythrins are most often used to store oxygen

Hemocoel The principle body cavity in molluscs and arthropods, remnant of the blastocoel. It forms part of the open circulatory system found in these animals. The true coelom is usually reduced to a cavity surrounding the heart, the pericardial cavity.

Hemolymph The fluid contained in the hemocoel of organisms with an open circulatory system. Hameolymph acts as both the coelomic and circulatory fluids.

Heterocercal The type of tail fin in a fish where the dorsal lobe is larger than the ventral lobe and the vertebrae of the axial skeleton form part of the dorsal lobe. This type of tail is typical of sharks.

Heterotherm Animals whose body temperature fluctuates. Fluctuations in body temperature are often related to temperature changes in the surrounding environment.

Heterotrophic Organisms not capable of converting light into chemical energy. They must consume other organisms or material produced by other organisms to survive. This is a characteristic of all animals.

Hindlimb The appendages attached to the posterior pelvic girdle of vertebrates. The hindlimbs may be modified as legs or fins.

Homeotherm Organisms that maintain their body temperature at stable temperatures independent of the surrounding environmental temperature.

Homology, Homologous Structures that have a similar evolutionary origin but different functions. The wing of a bat and a whale's flipper are homologous because both are derived from the anterior appendages—one is used for flying, the other for swimming.

Hydrostatic skeleton Formed from a fluid-filled and closed cavity surrounded by a body wall containing muscles oriented in different directions. Muscular contractions maintain the rigid form or change the shape of the organisms allowing movement.

Incomplete gut A digestive system that has only a mouth and no anal opening. Both ingested food and the undigested food must pass through the same opening to the alimentary tract.

Integument The body surface or covering of an animal.

Intercostal muscles Muscles between the ribs involved, in part, in breathing.

Intermediate host In parasites with complex life cycles involving more than one host, the organisms that contain the larval stages of the parasite.

Intertidal zone Part of the ocean ecosystem covered with water at high tide and exposed at low tide.

Introvert An eversible proboscis found in a number of the pseudocoelomate phyla. When the pharynx is everted, the hooks on the introvert are used to penetrate and hook onto the substrate or prey.

Jaw Made of either cartilage or bone, the jaw is a modified gill arch used for feeding. It may be armed with teeth or hardened plates, and it forms a part of the vertebrate mouth.

Kinetic skull Skull found in squamate reptiles, lizards and snakes, birds, and some fishes. Unlike other skulls, the upper jaw and palate that form the snout or nose can move relative to other parts of the skull.

Labium The second maxilla in some uniramians, such as insects, are often fused to form the lower portion of the buccal cavity.

Lacunae Empty spaces or cavities found between other cells that are otherwise solid or continuous.

Larval amplification A form of asexual reproduction where, during the life cycle, a single larval organism can produce large numbers of the next developmental stage. A good example is the fluke life cycle. A single sporocyst develops into hundreds of redia.

Leuconoid sponge The most complex of the three different sponge architectures. Choanocytes are found in chambers, and there is no spongocoel. Water enters incurrent canals to the prosopyles and exits the chambers through apopyles, then excurrent canals and the osculum.

Lophophore A unique double ring of hollow ciliated tentacles that surround the oral opening in a number of animal phyla. There is some debate as to whether these should be separate phyla, and many consider the lophophore to be a unique character of the phylum Lophophorata.

Lorica Any external protective casing that surrounds an invertebrate. Loricas are made of either materials secreted by the organism or surrounding materials cemented together to form the protective girdle.

Lunule In sand dollars, notches or holes that pass through the body of the animal. Their function is not fully understood but may be involved in movement of food from the oral to aboral surface, or they may stabilize the animal in strong water currents.

Macronucleus One of two types of dimorphic nuclei found in ciliate protozoans. The macronucleus contains multiple copies of the genome (polyploid) and is responsible for general protozoan cell function. The other type of nucleus is the micronucleus.

Madreporite The sievelike external opening to the water vascular system in echinoderms.

Malpighian tubules Excretory structures found in insects and some spiders. Although similar in appearance and function the two are not homologous. Malpighian tubules are hollow extensions of the gut suspended in the hemocoel.

Mammary gland The milk producing glands that, along with hairs, are characteristic of mammals.

Mandible The lower jaw of a vertrebrate or the feeding appendages of uniramians and Crustacea.

Mantle A thin sheetlike membranous extension of the visceral mass of molluscs that forms two flaps of skin. The mantle secretes the shell on the dorsal side and the space between the two flaps of skin is referred to as the mantle cavity.

Mantle cavity In molluscs the mantle, an extension of the body wall, secretes a shell creating a cavity between the mantle and the body of the animal.

Mastax The unique feeding structure of rotifers. It is a modification of the pharynx that includes internal jaws (trophi) used to grind and tear ingested food.

Maxilla Head appendages involved in feeding. They most commonly are involved in manipulating the food so that it can be crushed or chewed by the mandibles.

Maxilliped Arthropod appendages are referred to as maxillipeds when the first appendages found on the thorax become involved in feeding.

Medusa The free-swimming, mobile stage of the cnidarian life cycle. This stage, when present, is reproductive and mature gonads form on either male or female medusa. A common example is the jellyfish.

Mesenchyme The middle layer between the inner and outer cells of an animal with only two cell layers. It forms from ectoderm, consists of a jellylike mesoglea, and may also contain cells.

Mesocoel The middle of three coelomic spaces found in the tripartate body plan characteristic of the deuterostome lineage of animals. The other coelomic compartments are the protocoel and metacoel.

Mesoderm The third cell layer that develops in the gastrula between the ectoderm and endoderm in triploblastic animals. Mesoderm develops into muscle, connective tissues, and bones, as well as blood and other components of the vascular system.

Mesoglea The jellylike layer found between the ectodermal and endodermal cell layers of diploblastic organisms. It acts as a type of cement holding the two layers together but, unlike mesenchyme, has few, if any, cells.

Mesonephric Replaces the pronephros during development and is retained as the functional kidney in adult fish and amphibians. Formed posterior to the pronephros from renal tubules, which ultimately connect with the archinephric duct.

Mesosoma In animals where the abdomen (or in chelicerates, the opisthosoma) is divided into two regions, the most anterior is the mesosoma and the posterior the metasoma.

Mesosome In the tripartate coelom of deuterostomes, this is the middle of the three coelomic spaces.

Metabolic waste The biochemical waste products of metabolism. These most commonly include ammonia, carbon dioxide, and water.

Metachronal wave During the coordination of the ciliary movement, bands, or groups of cilia, are at different stages of their beating pattern, and this creates a wavelike appearance to the movement of the cilia on the surface of the organism.

Metacoel The last of three coelomic spaces found in the tripartate body plan characteristic of the deuterostome lineage of animals. The other coelomic compartments are the protocoel and mesocoel.

Metamere In segmented animals that have undergone metamerisation, each of the repeated units is a metamere. Each metamere contains identical structures to the metameres adjacent to it.

Metameric A body plan that consists of a series of identical units, metameres, repeated down the longitudinal axis of the animal. Each metamere contains identical structures to the adjacent metameres.

Metamerism The division of the body into a series of identical units, metameres, repeated down the longitudinal axis of the animal. Each metamere contains identical structures to the adjacent metameres.

Metamorphosis Distinct and marked changes between two stages in the life cycle of an organism. Examples are caterpillars and butterflies or tadpoles and frogs.

Metanephridia An excretory-osmoregulatory organ consisting of a ciliated funnel, nephrostome, connected to tubules that lead to the external nephridiopore. The nephrostome collects coelomic fluid, and all of its contents, to produce the urine.

Metasoma In animals where the abdomen (or in chelicerates, the opisthosoma) is divided into two regions, the more posterior is the metasoma and the anterior the mesosoma.

Metasome In the tripartate coelom of deuterostomes, the posterior of the three coelomic spaces.

Metazoan True multicellular organisms that exhibit all the characteristics of animals. They have cells, tissues, and organs.

Micronucleus One of two types of dimorphic nuclei found in ciliate protozoans. The single micronucleus contains only one copy of the genome and is used during the reproductive cell divisions. The other nuclear type is the macronucleus.

Microtrich Small microvillar projections on the surface of the tapeworm tegument. Michrotriches increase the surface area for exchange and may interdigitate with the microvilli of the host's intestine.

Microvilli Small, fingerlike projections of the cell surface that increase the surface area available for exchange.

Molting The periodic shedding of the outer body covering of an animal. The term is applied to a variety of things including: the loss of the outer exoskeleton in arthropods, the fur of a mammal, feathers of a bird, or the skin of a reptile.

Monociliate Cells that have flagella and, as a consequence, only one 9+2 organelle on each cell. This differs from polyciliate cells that have numerous cilia (9+2 structures) on the surface of each cell.

Monoecious Organisms that have, at some time during their life, both male and female reproductive structures. This is the opposite of dioecious.

Monophyletic A group of organisms, including the ancestor to that group, that all share a common evolutionary line of descent. (Compare to polyphyletic.)

Multiple fission A form of cell division where a single parent cell divides and produces more than two daughter cells. When only two daughter cells are produced, the process is referred to as binary fission.

Myoepithelium Epidermal cells specialized for contraction. Also called epitheliomuscular cells.

Myomere Blocks of segmental muscles found in chordates. The term may also be used to refer to similar blocks of muscle in other animals, but is traditionally used for chordates.

Myoneme Strands of contractile myofibers found in single cells. These allow the cell, or a portion of the cell, to contract in length and change its shape.

Myotome The embryological blocks of mesoderm in vertebrates that form the myomeres. In some cases myomere and myotome are used interchangeably.

Myriapod A taxon within the Unirmaia that includes animals that have many pairs of legs, but not those with three pairs—the insects.

Nacreous layer The innermost of the three layers of a mollusc shell. Also referred to as mother-of-pearl, it consists of thin layers or calcium carbonate crystals (aragonite), continually produced by the mantle surface.

Naupliar eye The unique eye found in the larval stage of the crustacean life cycle. This single compound eye is located medially, and with the exception of copepods, barnacles, ostracods and a few others, it disappears and is replaced with paired eyes in adults.

Nauplius larva A larval stage in the crustacean life cycle characterized by the single medial compound eye, the naupliar eye.

Nematocyst This organelle is part of the cnidocyte cell unique to the Cnidarians. It is the stinging, or eversible, portion of the cell, and it can drill into, entangle, and or stick to potential prey.

Nephridia A general term referring to the metanephridia and protonephridia.

Neutral buoyancy When the overall denisty of the organism is the same as the surrounding water. Body tissue is denser than water, and air or oils may be used to achieve neutral buoyancy.

Niche The role that a species has in its surrounding environment. The species' way of life and its combination with abiotoc and biotic defines that species and its role.

Notochord This ancestral characteristic of the Chordata, consists of a stiff cartilaginous rod near the dorsal surface of the animal. It is skeletal and helps support to the body.

Nutritive muscular cell The cells that form the gastrodermis lining the inner cavity of cnidarians. They carry out two functions: The first is to absorb and digest food and the second is in movement or changing shape by contraction of the myoneme portion of the cell.

Ocelli The simple eye of arthropods consisting of a single lens and optical unit. Ocelli are not usually image forming and detect levels of light instead.

Open circulatory system A circulatory system in which the circulating fluid (blood) flows into vessels or tubes not connected to each other by small capillaries but instead enters into the hemocoel before returning to the heart.

Opercular gill The gill found in bony fish. The operculum is the outer covering of the gill. When it is drawn away from the body, it pulls water inside the buccal cavity across the gills. This motion, in combination with the opening and closing of the mouth, aerates the fish gills.

Opisthaptor One of two attachment organs in monogenean flukes. Unlike the prohaptor, the opisthaptor is the principal attachment structure and can consist of hooks, claws, and suckers.

Opisthosoma The second tagma of the cheliciforme body, which either lacks appendages or, if they are present, involved in gas exchange or silk production.

Oral siphon Part of the body of an ascidian that surrounds the mouth. Water is pulled into the siphon chamber through the oral siphon by the action of the cilia lining the pharyngial basket.

Oral surface In radially symmetric animals there is no anterior, posterior, or right sides to the animal. Instead, we refer to the two sides of the animal by referring to the location of the mouth. In this case the side where the mouth is found.

Ossicle Either the small bonelike structures or plates that form the endodermal skeleton of echinoderms, the needlelike deposits of the shell in early molluscs, or the name of small bones found in the vertebrate inner ear.

Oviger legs The first pair of legs in male sea spiders (Pycnogonads) modified for holding and taking care of the eggs prior to their hatching.

Oviparous When the eggs are laid by the female and the embryo develops and hatches outside of the body of the female.

Paedomorphosis When sexually mature adults have characteristics that would normally be associated with the larval stage. It happens because reproductive structures form in the larval before the morphological change to the adult occurs.

Palp Small, sensory appendages, or feelers, associated with the feeding mouthparts of some invertebrates, most notably arthropods.

Parapodia Paired lateral, unjointed appendages of polychaete worms. Parapodia have a variety of shapes, and their appearance is related to their different roles in locomotion and respiration in different polychaete worms.

Parapodium One of the paired lateral, unjointed appendages of polychaete worms. Parapodia have a variety of shapes, and their appearance is related to their different roles in locomotion and respiration.

Parthenogenesis A form of asexual reproduction where viable offspring develop from unfertilized eggs that, depending on the organism, may be either haploid or diploid.

Pectine Modified appendages of the opisthosoma of a scorpion. Although their function is uncertain, they are thought to be mechanoreceptive sensing vibrations in the substrate.

Pectoral girdle Bones in vertebrates that connect the appendages on the left and right side of the anterior appendicular skeleton to each other. The pectoral girdles are also attached to the axial skeleton in amphibians, reptiles, birds, and mammals.

Pedicellaria Defensive pincerlike structures found on the surface of echinoderms. They are capable of responding to external stimuli independently from the main nervous system of the animal. Some complex pedicellaria are poisonous.

Pedipalp The second pair of appendages on the body of a cheliciforme arthropod.

Pelagic One of the zones in the marine environments where animals live. Pelagic animals are only found swimming or floating between the surface and bottom of the ocean and not at the shoreline.

Pellicle The network of semirigid cell membrane thickenings found on the surface of some protozoans. These are used to anchor either the locomotory flagella or cilia into the surrounding plasma membrane.

Pelvic girdle Bones in vertebrates that connect the appendages on the left and right side of the posterior appendicular skeleton to each other. The pelvic girdles are also attached to the axial skeleton in amphibians, reptiles, birds, and mammals.

Pentaradiate Radial symmetry based on five (from the Latin: penta for five). This type of symmetry is unique to the Echinodermata.

Pericardial cavity The part of the coelom that forms a sac, or space, that surrounds the heart. In invertebrates with a hemocoel, the pericardial cavity is all that remains of the true coleom.

Periostracum The outermost of the three layers of a mollusc shell. This horny layer consists of chonchin, a protein that protects the underlying layers from damage.

Peristomium The segment behind the prostomium in annelids that contains the oral opening. This segment like the prostomium is not a true segment and has no setal hairs, which are characteristic of the phylum. (Some textbooks erroneously refer to this as the first true segment.)

Perivisceral cavity The body cavity that surround the main organ systems in an animal.

Pesticide resistance When the target insects are able to survive a concentration of pesticide that previously was effective at controlling or killing them.

Pharyngeal gill slit Lateral opening in the wall of the pharynx that allows water to enter into the mouth and exit through the pharynx. This is one of the ancestral characteristics of the phylum Chordata. It allowed water to be removed from ingested food before it was passed back into the digestive system.

Pharynx The region of the digestive tract between the mouth and esophagus. In most animals it is muscular and forces food into the digestive tract that lies behind. it. In vertebrates it is part of both the digestive and respiratory tracts.

Phenetic taxonomy A method of classifying organisms based on numeric methods that evaluate organisms by a variety of measurements to create taxa without any consideration of phylogeny.

Phyllopodous In primitive crustacea the inner and outer surfaces of the leg are often enlarged to form flap, or leaflike structures that the crustacean uses for locomotion, feeding, and respiration.

Phylogeny The evolutionary history of either an organism or a taxon.

Pinnule In crinoid echinoderms the surface area of the arms is increased by smaller side branches, or extensions, referred to as pinnules.

Placoid scale Scales characteristic of the cartilaginous fish. Formed from the dermis, the scale is anchored in that layer by a basal plate composed of dentin and from that a spine, made of toothlike enamel, points backward.

Planula A planula is the solid, free-swimming larval stage of cnidarians consisting of two cell layers— an outer ciliated ectoderm and an inner endoderm.

Planuloid Something that resembles a planula, the solid, free-swimming larval stage of cnidarians consisting of two cell layers— an outer ciliated ectoderm and an inner endoderm.

Pneumatized bone Bone that contain numerous spaces and sinuses. Found in birds, these bones are light and strong due to internal struts and cross braces.

Polyp The sessile, asexual stage in the cnidarian life cycle. In some species they are independent organisms; in others, they form colonies where some polyps are involved in food gathering (gastrozooids) and other polyps produce the reproductive stage (gonozooids).

Polyphyletic A taxon that includes animals from two or more distinct evolutionary lineages and may not include the ancestor of either. (Compare to monophyletic.)

Precaudal The part of the body in front of the tail.

Primary host In parasites with complex life cycles that involve more than one host, the organism that contains the adult stage of the parasite.

Primitive characters The traits, or characteristics, that an animal has that are ancestral to the taxon.

Prismatic The middle of three layers in the typical molluscan shell.

Proboscis Tubular feeding structure that extends from an animal. In many invertebrates, it is eversible and extends from the body to permit feeding.

Proglottids Serially repeating segmentlike structures found in tapeworms that contain the reproductive organs. Immature proglottids are continually added at the scolex, and the most mature proglottids, containing thousands of eggs, are found the farthest from the scolex.

Prohaptor The specialized attachment sucker that surrounds the mouth in monogenean flukes. It consists of adhesive structures that help attach the animal to the host. The opisthaptor, not the prohaptor, is the principal attachment structure.

Pronephric The first, or ancestral, kidney that appears in the anterior part of the coelomic cavity and is connected to the archinephric duct. In amniotes and bony fish it appears only in the early stages of the embryo before it disappears.

Prosoma The first tagma of a cheliciform consisting of the first six segments of the body. Appendages on the prosoma are involved in feeding and locomotion. The prosoma has no sensory appendages such as antennae.

Prostomium The most anterior part of an annelid found just in front of the mouth it, like the peristomium, is not a true segment.

Protocoel The first of three coelomic spaces found in the tripartate body plan characteristic of the deuterostome lineage of animals. The other coelomic compartments are the mesocoel and metacoel.

Protonephridia An osmoregulatory-excretory structure found in some invertebrates. Also called a flame-cell, this tubule is closed at its distal end. The beating of internal cilia pull water across the cell membrane and then propels it down the tubule.

Protopodite The basal part of the crustacean biramous appendage that attaches it to the body. Consists of two parts: coxa and basis.

Protostome Phyla that share common characteristics of the blastopore forming the mouth, spiral determinate cellular cleavage in the embryo, and the formation of the body cavity by schizocoely.

Protozoans Single-celled organisms, in the kingdom Protista, that exhibit the animal-like characteristic of having to feed to obtain nutrients. They are considered heterotrophic.

Pseudocoelom A body cavity that is not completely lined by mesoderm. The mesoderm is associated with only the ectodermal surface but not the endoderm. There is some debate on how the psedocoelom forms. It is by definition a remnant of the blastocoel but often body cavities are defined as pseudocoeloms without properly making this developmental link.

Pseudocoelomate Animals that have a body cavity that is not completely lined by mesoderm. In the past these organisms were referred to as the phylum Aschelminthes but this is no longer considered acceptable because there is no apparent common ancestor to the group.

Pseudopodia A cytoplasmic extension that extends from the surface of either a protozoan or any amoeboid cell. These structures are temporary and are used for locomotion and feeding.

Pulmonary system That part of the circulatory system the transports blood from the heart to the lung, or other respiratory surface, and back to the heart. It's counterpart is the systemic system.

Quadraradial symmetry Body symmetry based on four identical parts arranged around a central axis.

Radial cleavage During development as the cells of the zygote divide, the products of the cell division remain stacked directly on top of each other.

Radial symmetry When an organism's body parts are arranged around the oral-aboral axis so that any plane passing through this axis results in two identical halves.

Radula This unique feeding structure is an ancestral characteristic of all animals in the phylum Mollusca. It looks like a tongue covered with teeth and works like a file to rasp food off the substrate.

Redia A stage in the life cycle of trematode flukes. The redia in the intermediate host develop from germ cell in the sporocyst. Redia also contain numerous germ cells that will develop in cercaria, an example of larval amplification.

Renette cell Unique cells found in nematodes believed to be involved in osmoregulation and elimination of metabolic wastes.

Respiratory tree The unique respiratory structure found in sea cucumbers (Echinodermata). Water is pumped in and out of these structures found inside the body cavity.

Rhabdite Found in the epithelium of free-living flatworms, and when released dissolve to produce mucous, which may also help in defense because it contains toxic chemicals.

Rhopalia These sensory structures are found around the bell margin of the jellyfish medusa. They always contain a statocyst for balance and sometimes an ocelli for light detection.

Rhynchocoel The unique body cavity of the Nemertean worms, which contains the retractable proboscis. There is some debate on whether this is a true coelom, making it difficult to state whether the Nemertea are eucolomates.

Scalid Rings of spines at the base of the head directed toward the posterior end of the body of Kinorhynca and Loricifera.

Schistosomiasis A disease cause by the trematode parasite Schistosoma found around the world. Different species are responsible for diseases ranging from common swimmers itch to schistosomiasis, which effects over 200 million people.

Schizocoel A true body cavity that forms by schizocoelus splitting of the mesoderm.

Schizocoelus One of two ways that a body cavity forms within the mesoderm. (The other is enterocoelic pouching.) The block of mesoderm splits apart to create the coelomic space, a charactersitic of the protostomes.

Schizogony A form of asexual reproduction found in some protozoans. An already multinucleated cell undergoes cell division that results in each daughter cell containing only one of the many nuclei present in the parent cell. This is also referred to as multiple fission.

Sclerotised Chemical covalent crosslinking of separate protein chains using phenolic compounds. Sclerotised protein is stable and is not easily broken down or digested.

Scolex This unique attachment organ of the tapeworms is the most anterior part of the animal and is used to attach to the host. It consists of adhesive suckers and, in some species, hooks. Proglottids develop from behind the scolex.

Scyphostome The jellyfish, class Scyphozoa, get their name from this unique stage in the life cycle. It is a small polyp existing for only for a short time before developing into a strobila, which then produces the medusa.

Segmentation The division of the body into a series of identical segments, metameres, that are repeated down the longitudinal axis of the animal. Each metamere contains identical structures to the metameres adjacent.

Semicircular canal Part of the inner ear consisting of three semicircular ducts that are arranged at right angles to each other. They are sensitive to angular acceleration of the head in the three different axes of space.

Septum, Septa Sheets of tissue that separate two compartments or cavities. Plural is septa.

Serial homology Metamerization results in a linear series of segments that share a common embryonic origin. Ancestrally, all metameres were identical. Modification of different parts of the body for different functions means that this similarity has been lost.

Setae Bristles, or hairlike structures, that extend from the body. Usually made of chitin, they are common in annelids and arthropods.

Siphonoglyph A pair of ciliated grooves in the oral opening of anthozoans that, along with other paired features, create the biradial symmetry of the group. The cilia propel water into the gastrovascular cavity.

Siphuncle A cord of tissue that passes between the different chambers of a Nautilid shell. The siphuncle regulates the amount of air or fluid in the chambers to assure neutral buoyancy for the animal as it swims.

Skeletal girdles In vertebrates, bones that connect the appendages on the left and right side of the appendicular skeleton to each other. The skeletal girdles are also attached to the axial skeleton in amphibians, reptiles, birds, and mammals.

Spicules Any needlelike structure. This term is most often thought of in conjunction with sponges and refers to the needlelike structures produced by sponge cells that form the supporting skeleton. In Nematodes spicules are used by the male during copulation. Needlelike deposits of the shell in Mollusca are also referred to as spicules.

Spinneret The modified appendages on the posterior end of the opisthosoma of a spider that produce silk.

Spiral cleavage Pattern of cell division in the developing embryo where the products of the cell divisions shift by rotating either clockwise or counterclockwise so that the resulting daughter cells lie in the furrow of the underlying pair of cell. The opposite of radial cleavage.

Spiral valve In sharks, a modification of the digestive tract to slow the movement of food and to increase the surface area. In an amphibian, it helps to keep systemic and pulmonary blood separate in the single ventricle of the heart.

Spongocoel The internal cavity of asconoid and syconoid sponges that opens to the outside through the osculum. There is no spongocoel in a leuconoid sponge.

Sporocyst A stage in the life cycle of trematode flukes. The sporocyst develops from the mericidium found in the intermediate host. Each sporocyst contains the germ cells that will develop into numerous redia, an example of larval amplification.

Sporogony A form of asexual reproduction where the fusion product of the male and female gamete (the zygote) undergoes multiple cell divisions that produce sporozites. Found in animals in the protozoan phylum Apicomplexa.

Statocyst A balance organ that senses gravity. It consists of at least one solid statolith surrounded by sensory cilia. As the position of the organism changes, the statolith rolls stimulating different cilia.

Stolon In some colonial cnidarians, ascidians, and lophophorans, a cord of tissue connects different zooids of the colony to each other. Buds on the stolon give rise to new zooids, which retain their connection to the rest of the colony through the stolon.

Strobilization The process that converts the scyphostome into a strobila during the life cycle of schyphozoans, jellyfish. Transverse divisions of the strobila produce the small, disc-shaped ephyra that develop into the adult jellyfish.

Suctorial feeders In this type of feeding strategy food is sucked into the digestive tract by the muscular action of the pharynx.

Swim bladder Found in bony fish, this gas-filled chamber is used to main neutral buoyancy. Oxygen in the blood is added or removed as required. The swim bladder in some fishes opens into the digestive system allowing these fish to swallow air instead.

Syconoid sponge Of the different sponge architectures, this is intermediate in its complexity. The spongocoel is no longer lined with choanocytes, now located in radial canals that open to the spongocoel through apopyles. Water enters the radial canals through prosopyles and exits through a single osculum.

Synapomorphy A new and unique character that a group of organisms all share and that defines the lineage or clade.

Synapsid Vertebrate skull of the mammals and the extinct mammal-like reptiles. There is only one temporal opening behind the orbits in the skull.

Syncytial Protoplasm that contains numerous nuclei not separated from each other by plasma membrane. This creates a multinucleate cellular appearance for a tissue that appears to lack cell boundaries.

Systemic system That part of the circulatory system the transports blood from the heart out to the body and back to the heart. It's counterpart is the pulmonary system.

Tagma, Tagmosis, Tagmatization The distinct body regions resulting when different segments of a metameric animal become involved in specific functions. These segments are modified to carry out that function and their appearance changes.

Taxon Any taxonomic group of related organisms. Plural is taxa.

Tegument The outer covering of parasitic flatworms, including flukes and tapeworms, consisting of a syncytial outer layer of cytoplasm connected to cell bodies embedded deep in the underlying protective mesenchyme.

Telson In a number of arthropods the most posterior part of the body, which may appear as a tagma, is not a true segment. This part of the body, formed from the embryonic pygidium, is referred to as a telson.

Test An outer, nonliving shell or case that surrounds an organism.

Tetrapod Vertebrates that have four limbs. The forward pair is attached to the pectoral girdle and the posterior pair to the pelvic girdles. Tetrapods include amphibians, reptiles, birds and mammals.

Thorax In organisms that have undergone tagmosis the trunk often develops into two tagma, one involved in locomotion and one that is not locomotary. The locomotary tagma is referred to as a the thorax.

Torsion An unusual twisting of the gastropod body that has left all members of the class with an asymmetric body plan and a U-shaped alimentary tract, with the normally posterior anus and mantle cavity now located anteriorly over the head.

Totipotent When a differentiated cell can change into any one of a variety of different specialized cells found in an organism. These changes will result in the cell carrying out new functions.

Trachea Part of the respiratory system in insects, vertebrates, and some spiders. In the invertebrates it consists of tubules carrying air directly from outside to the tissues. In vertebrates trachea are tubes moving air to the lungs.

Tracheal system The respiratory system found in insects and some spiders where air passes directly to the tissues through a series of tubules. Because oxygen transport is direct, respiratory pigments are often missing from the blood of animals that use a tracheal system.

Tripartate coelom Characteristic of the deuterostome line of animals. The coelom forms with three separate compartments referred to as the protocoel, mesocoel, and metacoel.

Triploblastic Organisms formed from the three cell layers: endoderm, ectoderm, and mesoderm.

Triradiate Radial symmetry based on three.

Trocophore This free-swimming ciliated larval stage is found in a number of animal phyla including the Mollusca and Annelida. The larvae has a unique circle of preoral cilia around the middle of the body. Trocophores are often considered an ancestral characteristic of protostomes.

Trophi The paired chitinous jaws found inside the mastax of a rotifer. The trophi are adapted to the feeding strategy of the rotifer and can grind, cut, or be used to capture prey.

Trunk In animals with only two tagma, the part of the body behind the head, or first tagma.

Tube feet Hollow and fluid-filled tubes that are part of the water vascular system in Echinoderms. Muscles associated with the tube feet allow them to be hydraulically controlled and function in locomotion, attachment, food gathering, and gas exchange.

Tunic The outer covering of the tunicates secreted by the underlying body wall.

Typhlosole This invagination of the gut wall in earthworms increases the surface area of the gut available for digestion.

Uniramous A type of appendage in arthropods. Uniramous appendages are unbranched and each element of the leg is arranged in a linear sequence.

Vector The animal (often an insect) carries a disease from an infected individual to one that is not. For example, the mosquito is the vector of malaria in humans.

Velum A thin flap of tissue found around the inner surface of the bell of a hydrozoan medusa.

Ventricle Chamber of the heart that pumps the blood out and away from the heart.

Vermiform Animals with a wormlike appearance with a cylindrical body, considerably longer than it is wide.

Visceral mass One of three parts of a mollusc: head, foot, and visceral mass. It consists of the fleshy part of the mollusc and contains the main organ systems of the mollusc.

Water expulsion vesicle Special organelles found in protozoan and some parazoans that is involved in osmoregulation. The organelle collects water from the cytoplasm and then releases it from the cell. They are also called contractile vacuoles.

Water vascular system A characteristic of echinoderms. It is a modification of the coelom, and this closed, water-filled system forms canals and branches throughout the body. In one part, the tube feet, it acts as hydrostatic skeleton permitting locomotion.

Zoaria A colony of lophophorate ectoprocts (bryozoans).

Zoecium The nonliving outer exoskeleton of an ectoproct (bryozoan) colony.

Zonite The name for the thirteen segments of a kinorhynch. The segments are triangular. Rings of spines at the edges of the zonite help in movement.

Zooid In colonial animals, each member of the colony. Often there is a division of labor between zooids, the most common being between feeding gastrozooids and reproduction gonozooids.

Notes

Notes

Notes

Notes

Notes

Notes